DECISION ASSURANCE

DECISION ASSURANCE

How to Make Decisions You Can Prove Were Right

Andrew Guitarte

Arclave, Inc.

Decision Assurance: How to Make Decisions You Can Prove Were Right

Published by Arclave, Inc.

ISBN (paperback): 979-8-9968047-0-2

ISBN (hardcover): 979-8-9968047-1-9

First edition, 2026.

Cover and interior design by Arclave, Inc.

www.arclave.com

For Rizza — my wife, my lighthouse.

And for Ralph, Zach, and Blaise —
my boys, my inspiration, who keep me honest
the way only the young and innocent can.

“Somebody must audit the auditor.”

Contents

About This Series

A word on where this book sits, because it belongs to a longer work.

Behind *Decision Assurance* stands a larger idea, one I call **Emergent Intelligence.** It is simple to state and hard to exhaust: the most complex problems a civilization faces rarely yield to one more layer of central control. They yield, when they yield at all, to many small, well-governed parts acting on a few clear rules — the way an ant colony solves problems no single ant could grasp, or a market sets a price no single trader could name. Intelligence, in this view, is not a thing you install at the top. It is a pattern that *emerges* from the parts when the parts are sound. Emergent Intelligence is the study of how to build systems — and institutions, and disciplines — that decide and adapt that way.

Decision Assurance is the **first applied discipline** under that idea. It takes one wicked problem — how an organization decides, and how it proves it decided well — and turns it into a discipline you can practice on a Monday. It comes first on purpose, because nearly every other hard problem an institution faces sits downstream of the same gap: we never learned to assure the decision itself.

Here is the shape of the work, so you can see where this volume falls:

The LeanIX Payoff — the practitioner's foundation: turning an enterprise-architecture investment into demonstrable business value. The ground this all stands on.

***Decision Assurance** — this book.* The first applied discipline: making decisions you can prove were made well.

Emergent Intelligence — the forthcoming capstone: the theory itself, stated in full — the arc its author spent twenty-five years building toward.

And the disciplines that follow — because a single discipline is a beginning, not a library.

What holds the series together is not a list of topics. It is a single through-line: **each volume is a discipline for deciding well in a complex world.** That is broad enough to grow and specific enough to mean something.

And it is built the way the discipline itself prescribes — in the open, practitioner to practitioner, as a shared body of knowledge owned by no one and in service of everyone. If that sounds like the book you are about to read, that is the point. The method and the movement are the same shape.

Welcome to the Library.

The Decision Assurance Manifesto

We are drowning in decisions.

Every hour, in every team, people make calls that shape the enterprise — and the world. Most of those calls vanish the moment they are made. No record of who decided. No trace of why. No proof the decision was sound.

For our products, we fixed this long ago. Factories built Quality Assurance. Software built it too. The most expensive thing any organization makes — its decisions — earned nothing.

Now Artificial Intelligence (AI) pours fuel on the fire. It multiplies our decisions a thousandfold and whispers answers we cannot check. It can be confident and wrong in the very same breath. Speed without proof is not progress. It is risk, compounding by the day.

We refuse to accept that. A decision you cannot defend is a decision you do not truly own. And there is a better way.

What we value

We still prize speed, judgment, and human wisdom. We simply refuse to lose the proof.

We value decisions we can **prove** over decisions we merely **trust**.

We value reasoning **on the record** over reasoning locked in one person's memory.

We value standards the **system enforces** over standards that lean on vigilance.

We value **assured human judgment** over automated judgment that answers to no one.

That last line matters most. Decision Assurance never takes the decision away from you. You decide. It proves you decided well.

What we hold to be true

Every consequential decision deserves a record — who decided, why, what was weighed, and what was chosen.

The rigor of your assurance should match the cost of being wrong. Aviation grades its software that way. Our decisions deserve the same.

The rules that must never break belong in the system, not in someone's willpower.

A claim with no evidence is a guess. Ground every decision in something a stranger could check.

Dissent is data. Record the objection, not only the verdict.

Memory is an asset. A decision you can find, learn from, and defend pays you back for years.

Assurance serves everyone — the maker, the auditor, the customer, and the public.

Proof is not bureaucracy. Done well, it sets people free to decide with confidence.

The call

Act now.

Every day we wait, the risk climbs higher — technical debt, obsolescence, and a quiet backlog of decisions no one can defend. AI grows more powerful every week, and with it grows the danger: hallucination, opacity, and choices no one can explain, audit, or answer for.

Decision Assurance stops the rot. It brings transparency where there was fog, accountability where there was none, and proof where there was only hope. It is how we keep fairness and ethics in view instead of taking them on faith.

Do not wait for the audit, the failure, or the regret. Build assurance into the way you decide — now, while it is still a choice and not yet a crisis.

Somebody must audit the auditor. Let it be us.

Join us. Architect it. Build it. Prove it.

Founding signatory: Andrew Guitarte — and every practitioner who refuses to decide in the dark.

How to Use This Book

Before we start the walk, a short word on the three different ways to take it — because this book is built to be more than read.

It is, first, an argument: a case made cover to cover, each chapter opening onto the next, meant to be walked straight through at least once. If that is all you ever do with it, it will do its job. But two other readers are built into these pages, and if you are one of them, a little orientation now will save you time later.

If you are reading alone — the practitioner. Read it like a walk in a park: one chapter at a time, in order, no rush. Two features turn the reading into doing. Every chapter ends with a **"So what?"** (why this matters to you, in your chair) and a **"What now?"** (one concrete thing to do before the week is out). Treat the *What now?* as a standing invitation, not a suggestion — the discipline in this book is built one real decision at a time, and the first one is always the hardest to start. The end-of-chapter **key takeaways** are your review; the **key terms** are your working vocabulary; the **Glossary** at the back is the full lexicon for the moment a word slips.

If you are reading with others — the team. This book was built to be run by a group, a chapter a week. The end-of-chapter apparatus is the meeting agenda, already written for you: open with the **key takeaways**, settle the vocabulary with the **key terms**, test the room with **check your understanding** (the answer key is printed so no one needs to be the teacher), then spend most of the hour on the **questions for reflection and discussion**, which are written to surface how *your* organization actually decides. Close with the **apply it** exercise — and here is the move that turns a reading group into a working one: do not only discuss the chapter, *assure a real decision* with it. A team that walks one live decision through the framework together, week by week, finishes the book with a working practice, not just a shared opinion.

If you are teaching it — the instructor. Every chapter is already a lesson. The six-part apparatus — takeaways, key terms, check-your-understanding with an answer key, discussion questions, and a hands-on *apply it* — is a train-the-trainer kit, built so someone a single chapter ahead of the room can still teach it well. The back matter is your reference shelf: **Appendix A** is the formal specification of the Decision Assurance Levels, liftable straight onto a slide; **Appendix B** is the evidence base, for the student who wants to argue with the conclusions; **Appendix C** walks one decision through the entire discipline, end to end, as a capstone worked example; the **Field Kit** collects the checklists; the **Glossary** is the vocabulary list; and the **References** are the syllabus behind the syllabus. A semester, a workshop, or a lunch-and-learn series can be built directly on this spine.

One thing all three readers share. The book is also a **handbook**, which means you are meant to write in it, dog-ear it, and return to it — not finish it and shelve it. When a real decision lands on your desk eight months from now and someone asks you to prove it was sound, the chapter you need is the one you will reread, and the checklist you need is in the Field Kit. A handbook earns its keep on the second visit.

Now — whichever reader you are — turn the page. The walk starts with the size of the problem, and it is bigger, and stranger, than almost anyone admits.

INTRODUCTION

The Decision You Cannot Defend

Let me tell you why I will not let this go.

In 2016 I worked for one of the largest banks in the United States. I was a business architect — the person whose whole job is to see the enterprise as one thing: to know what the right hand is doing while the left hand moves, to hold the big picture when everyone else holds a corner of it. If anyone in that building stood a chance of seeing the trouble coming, it was someone in my seat.

One afternoon I sat in the cafeteria of one of our office buildings and looked up at a television on the wall. Our leadership was being grilled on national news. The story by then was everywhere, and it was simple and devastating: over years, millions of accounts were opened in customers' names without their permission. The regulators came. The fines climbed into the billions before it was over. People I will not name lost their jobs and their reputations. The institution I grew to love failed the very people it existed to serve — and some of those people were my own neighbors, my own family, me.

I want to be careful about what I am saying, because these facts are a matter of public record and I will keep them there. I am not here to relitigate a company or to point a finger at anyone inside it. The truth is gentler than blame, and more frightening. It was not a building full of bad people. It was a building full of good ones — talented architects, careful auditors, a wall of dashboards — and the failure happened anyway, quietly, for years, in the space between all of them. No one person was watching the whole, because no system made it anyone's job to.

That is the part that lit a fire in me, and the fire still burns. I held the widest lens in the building, and the lens was not enough. Judgment was not enough. Good intentions were not enough. Dashboards were not enough. The dots were all there; nothing connected them, and nothing made connecting them anyone's job. I stood in that cafeteria and made myself a promise of two words. *Never again.* Not on my watch, not in my profession, not while I could build the thing that would catch it.

This book is that promise, kept. Decision Assurance is not an idea I admired from a distance in a journal. It is what I went looking for the day I understood that loving an institution does not protect it, and being its finest architect does not protect it either — that the only thing that protects it is a system that watches the whole, on the record, before the quiet failure becomes the public one. I never want another person to look up at a television in a cafeteria and feel what I felt that afternoon. So we build the assurance. We do not sleep on the job. Never again.

You do not need to witness a national scandal to feel what I felt that day. The same gap opens on an ordinary Tuesday — smaller, quieter, and far more common. Let me show you the everyday version.

Picture a Tuesday afternoon.

You are three meetings deep. Somewhere between the second coffee and a message thread you keep meaning to answer, a note lands from someone in Audit. It is perfectly polite. It is also a small grenade.

Can you point me to the rationale for the platform decision from back in March? Who approved it, what options got weighed, and why did we choose the path we chose?

You remember the decision. You were in the room. You remember that it mattered. So you go looking for the *why* — and you find a one-line title in a change ticket, a half-dead thread in a tool nobody opens anymore, and a calendar invite for a meeting whose notes never quite got written. The decision happened. The proof did not.

If that little scene makes you wince, even a bit, then you and I are going to get along. This book is for you.

Let me be honest about something from the first page: I wrote it because I got fed up. Fed up with the delays and the shrugs. Fed up with decisions that vanish the moment they are made, with rationale that lives only in someone's memory, with the quiet dread that arrives the day a competent person asks you to prove a call was sound — and you simply cannot. We sent people to space. We sequenced the genome. We taught machines to talk. And we still cannot reliably answer a child's question about our own choices: *why did you do that?*

It is not rocket science. And yet, here we are.

You are not bad at deciding. You are drowning.

Let me say the kind thing first, because it is also the true thing: the problem is not you.

Think about the sheer volume. Your day begins with decisions before you are even awake enough to call them that — snooze or rise, coffee or run, answer the message or let it wait. By the time you reach the office, real or virtual, your brain already paid a small tax a hundred times over. Psychologists name the late-afternoon version of this *decision fatigue*, and you feel it most right when the biggest calls tend to land.

Now multiply your private little flood by everyone in the building. Every salesperson reading a customer. Every engineer choosing a shortcut or refusing one. Every manager deciding to escalate or let it ride. Thousands of decisions an hour — most of them small, some of them quietly enormous, and almost none of them written down in a way you could ever find again.

Then we handed the firehose a bigger nozzle. Artificial Intelligence (AI) now drafts our options, summarizes our meetings, and proposes our answers — faster than any of us, and with a confidence that does not flinch even when it is wrong. That is the part that should keep us up at night: a tool that multiplies our decisions a thousandfold and cannot always tell us which of its answers is real.

So no, you are not bad at this. You are a capable person standing under a waterfall, asked to remember every drop.

The questions nobody can answer

Here is the tell. Listen for these in your own week, because they are everywhere:

Do we already own a decision on this? Who decided? Why that option and not the others? Where is it written down? When was it final? How did we even get here? Are you sure?

Every one of those is reasonable. Every one of those deserves a clean answer. And most of the time, in most organizations, the honest reply is a slow blink and a search bar.

I sat through years of these moments from the inside. I chaired and attended more review boards than I can count — change boards, architecture boards, the rooms where a company decides what it will build and what it will retire. So let me share the open secret: a great many of those decisions get rubber-stamped. Someone reads a title. Someone asks, "any objections?" The silence gets recorded as wisdom. "Approved." Thank you. You're welcome. Next.

A dozen of those an hour, across an enterprise, every working day — each one touching real systems and real money. And almost none of them revisited to ask the only questions that ever teach you anything: *Was it the right call? What actually happened? What would we do differently?*

We do not track that. We barely track the decision itself.

We solved this once already

Now for the strange, slightly embarrassing part.

We already know how to fix this. We worked it out more than a century ago — just not for decisions.

Walk into any factory that makes something you would trust your life to, and you will meet two kinds of people doing two different jobs. One builds the *system* that makes defects nearly impossible: the checklists, the calibrated machines, the standards. The other inspects what comes off the line and catches the bad unit before it reaches you. We gave those jobs names — Quality Assurance and Quality Control — and no serious company would dream of running without both.

We built this discipline for our products. We built it for our software, our food, our buildings, our medicines. We never built it for the most expensive thing our organizations produce — the decisions themselves.

Closing that gap is the whole reason for this book. There is a name for it, and the name is **Decision Assurance**.

What Decision Assurance is — and what it refuses to be

Before we go further, here are the two definitions this book plants in the ground — the first formal definition of **Decision Assurance**, and its established counterpart, **Decision Control**:

> ***Decision Assurance*** *is the discipline of systematically ensuring that a decision is made well — completely, consistently, within its governing rules, and on real evidence — and recording it so the decision can be proven and defended afterward, with accountability always resting on a human or an organization, whether the decision is made by a person, an AI, or the two together.*

> ***Decision Control*** *is the discipline of judging an individual decision — ratifying or rejecting it at the gate and monitoring it thereafter — so a flawed decision is caught at or near the moment it is made. It is the reactive, decision-by-decision counterpart to Decision Assurance, and the established term in agency theory (Fama and Jensen, 1983).*

That is the formal language, for the record and the citation. Now the human version, the one I would give you over coffee.

Decision Assurance is the discipline of making sure a decision is made *well* — and being able to prove it later. Not "well" in some misty, inspirational sense. Well in a way you can show a regulator, an auditor, a board, or your own future self: the right things got considered, nothing important got missed, the rules that must never break did not break, the reasoning rested on real evidence, the dissent got heard, and the whole thing left a trail.

Notice what that is *not*.

Decision Assurance is not a robot that decides for you. This is the line I will defend in every chapter, so I will plant it right here at the start: a decision-assurance system does not make the decision. It makes the decision *defensible*. You still decide. You still own the call. The system simply guarantees that you saw everything that mattered, applied the same standards you would want applied every time, and left proof that you did.

In a moment when every vendor on earth is promising a machine that will think for you, that distinction is not a small one. It is everything. The winners will not be the ones who hand their judgment to a confident black box. They will be the ones who keep the judgment where accountability belongs — and *assure* it with everything technology can now offer.

Why me, and why now

A fair question: why listen to me about this?

For twenty-five years I built bridges between the business and the technology meant to serve it. I led the transformations, sat on the boards, wrote the strategies, and felt — personally, more than once — the cold drop of being unable to defend a decision I was certain was right. I also did something most people who write confident books about AI never do: I ran the experiment on myself. I built a system to test whether a board of AI agents could out-decide its own best single, well-calibrated agent — not a human expert, but the strongest member of the board itself. I pre-registered my hypothesis — the way scientists do when they want to be held to account — and then I let the results land wherever they would. I will tell you that whole story later, including the part where my favorite idea did not survive contact with the data. That story is the reason I trust this discipline; it earned that trust against my own hopes.

As for *why now* — look at the calendar, then look at the technology. AI grows more capable every single week, and with that power comes a rising tide of risk we are not ready for: confident hallucination, decisions no one can explain, accountability that evaporates the moment something goes wrong, and fairness and ethics we take on faith because we cannot actually watch them. Every month we wait, the backlog of undefendable decisions grows, the technical debt compounds, and the reckoning moves closer. There is a line I keep returning to: *somebody must audit the auditor*. I think that somebody is us, and I think the time is now — while this is still a choice and not yet a crisis.

How to read this book — and the promise I make you

This is not a textbook, and I did not write it like one. I wrote it the way I would walk you through a park: one delight at a time, each turn opening onto the next, nothing rushed and nothing wasted. If a chapter ever makes you feel talked down to, I failed — and you should tell me. You bring years of hard-won judgment to these pages; my job is to be useful to *that* person, not to lecture a beginner.

So here is the shape of our walk.

First, we sit with the problem squarely — how big it is, why the "smarter machine" keeps seducing us, and what I learned when I tested that seduction against reality. Then we draw the one distinction that makes everything else click into place: assurance versus control, the proactive system versus the reactive gate. From there we build the framework — the rules, the levels, the earned authority, the memory — the actual machinery of assuring a decision. Then we get practical: how you stand this up on a Monday morning, where the human stays firmly in the loop, and how the idea reaches far past architecture into any decision that matters. And we close by looking at the horizon — at what all of this means when the decisions in question belong not to a company but to a civilization.

Every chapter answers two questions, on purpose. *So what?* — why this matters to you, in your chair, with your name on the call. And *what now?* — one concrete thing you can do about it before the week is out. Later, I even lay out what an ideal decision-assurance tool should do — the gold standard, measured against the highest bar I can describe — and then I hold my own work up to that same bar and show you, plainly, where it shines and where it still falls short. You deserve nothing less than that honesty, especially from someone with something to sell.

That is the deal I am offering you. No jargon for its own sake. No hand-waving. No pretending the hard parts are easy. Just a straight, warm, occasionally funny conversation between two people who care about getting decisions right — and who are tired of being unable to prove that we did.

The audit email is still sitting in your inbox. Let us go figure out how you will answer it — this time, and every time after.

Turn the page. The first stop on the walk is the size of the problem, and it is bigger, and stranger, than almost anyone admits.

The Introduction, in review

The book doubles as a course and a handbook. The front and back matter carry a light version of the end-of-chapter kit; the full workbook begins with Chapter 1.

Key takeaways

The most expensive thing an enterprise produces is its **decisions**, and almost no one measures, assures, or can later defend them. That gap is the book's subject.

You are not bad at deciding. You are **drowning** — the volume is inhuman, and Artificial Intelligence (AI) is multiplying it.

The symptom is a set of reasonable questions no one can answer: *who decided, why, on what evidence, where is it written?*

We built assurance for our products, our software, and our money — never for our decisions.

The discipline that closes the gap is **Decision Assurance**: making a decision well, and being able to prove it later. Its established counterpart is **Decision Control**, the verdict at the gate.

Decision Assurance does **not** make the decision for you. It makes the decision *defensible*. Accountability stays human — even when an AI does the deciding.

This book is a handbook: every chapter asks *so what?* (why it matters to you) and *what now?* (one thing to do), and it judges its own author's tool against an honest bar later on.

Key terms

Decision Assurance — the discipline of ensuring, and being able to prove, that a decision was made well; the Quality Assurance of decisions.

Decision Control — the verdict on a specific decision, ratify or reject at the gate; the Quality Control of decisions.

Decision fatigue — the depletion that comes from making too many choices, so judgment degrades as the day wears on.

Accountability (versus authorship) — answerability for a decision, which never transfers to a machine.

A question to carry in

Recall your own "audit email" — a time someone asked you to justify a past decision, and the rationale was gone. What did you wish you could produce? Keep that moment in mind. The rest of the book is the answer to it.

PART I

The Problem

CHAPTER 1

The $2.3 Trillion Blind Spot

Let me take you onto a factory floor for a minute, because they worked out something the rest of us never did.

On any floor that makes a thing that could hurt you if it failed — a brake, a syringe, a part of a jetliner — you will find the place measured to within an inch of its life. Sensors on the machines. Tolerances on the parts. A person who pulls the tenth unit off the line and tests it until it breaks. Another person with a clipboard who watches the whole process so the defect never forms in the first place. Every input counted, every output checked, every step on the record.

It was not always this way. The factory earned that discipline the hard way, across a long century of being burned. Quality was once a matter of pride and luck — a good foreman, a careful crew, and fingers crossed. Then the defects, the recalls, and the harm piled up, and a handful of stubborn people — Walter Shewhart, W. Edwards Deming, the engineers at Toyota — insisted that quality is not an accident you hope for but a system you build. They measured the thing to death because they decided, after enough pain, that the cost of being wrong was too high to leave to hope.

Software learned the same lesson on the same schedule. We shipped, it broke, we shipped again, until testing, code review, and the discipline we now call site reliability turned "it works on my machine" into something a stranger could trust. Money learned it earliest of all: the entire profession of audit exists so that a number on a page can be believed by someone who was nowhere near the room when it was written. And in the most unforgiving corner of engineering — the software that flies an aircraft — the rigor goes all the way to a standard that grades each component by what happens if it fails, and demands the most punishing scrutiny for the parts whose failure would be catastrophic. Every one of these fields arrived at the same conviction by the same painful road: the things that matter most must be assured, not merely attempted.

Now walk into the room where your company makes its most expensive thing, and count the instruments.

There are none.

The most expensive thing your enterprise produces is not a part or a product. It is a *decision* — to fund this, to kill that, to buy, to build, to migrate, to wait. Those decisions commit the budget, set the direction, and live with you for years. They are the output every other system in the building exists to serve. And almost nowhere do we measure them, instrument them, or assure them the way that factory assures a brake. That is the blind spot. It is the largest unmeasured surface in the modern enterprise, and we walk past it every single day.

You will notice a number on this chapter's title — $2.3 trillion — and you deserve to know early that it is not a scare figure I made up. It is a real, published estimate, and in a few pages I will show you precisely where it comes from and why it is almost certainly an undercount. For now, hold it loosely — a marker of how large the blast radius gets when decisions go unassured at scale. First, let me show you the gap itself.

We measure everything else

Sit with how strange that is, because we are not careless people. The same enterprise that cannot tell you why it funded a project will instrument a web page down to the millisecond. We know our net promoter score to a decimal place. We track page-load time, cost-per-click, inventory turns, defects per million, mean time to recovery. We run experiments on the color of a button and argue about the results with real data. We assure our code, our supply chains, our financials, our food, our buildings. We built entire professions — quality assurance, audit, safety engineering — to make sure the important things are done right and can be proven later.

Then ask the one question that should stop a meeting cold: *what is our decision accuracy?* What fraction of last year's funded bets actually paid off? How often does this board approve something it later wishes it could take back? How long, on average, does a proposal sit before it gets a yes or a no? Nobody holds the number. Not because the question is unreasonable — it is the most reasonable question in the building — but because the data to answer it was never kept.

Think about a decision your company makes constantly: build it ourselves, or buy it from a vendor. That single call, repeated across hundreds of projects, steers enormous sums and shapes the architecture for a decade. Now try to find, anywhere, a scorecard of how those build-versus-buy calls turned out. You will not find one. We measure the inputs to our decisions and the outputs of our decisions, and we leave the decision itself — the hinge in the middle — completely dark.

How big is the bill?

Honest answer first: no one can hand you a clean total, and that is the whole point — we do not count decisions, so we cannot add them up. What we can see is the blast radius.

McKinsey surveyed more than 1,200 leaders and pinned the cost of inefficient decision-making at roughly 530,000 days of managers' time a year for a single large company — on the order of $250 million in salary, burned annually, by one firm, on deciding slowly and deciding badly. Widen the lens and the figures turn vertiginous: by one estimate, around $2.3 trillion of digital-transformation spending worldwide poured into programs that failed — and behind a failed transformation you almost always find a trail of decisions no one could defend, revisit, or learn from. Pick whichever number you can stand to look at. The order of magnitude is not in dispute.

But the dollar figures, large as they are, undercount the real damage — because decisions compound. A product is finished when it ships. A decision is never finished; it becomes the foundation the next decision stands on. Approve the wrong platform, and every system built on it for the next five years inherits the mistake. Bury the reasoning behind a reorganization, and the next reorganization repeats it, because no one could discover why the last one failed.

Engineers gave a name to the silent interest you pay on shortcuts in code: technical debt. There is a quieter, costlier cousin, and the whole enterprise carries it without a line item — *decision debt.* Every choice made with no record, no rationale, and no way to learn from it is a loan taken against the future. And like all debt left in the dark, it does not sit still. It compounds, in silence, until the day it comes due all at once.

Why we let it happen

Why did we let this happen, disciplined as we are everywhere else? Two reasons, and they feed each other.

The first: a decision hides its defects. A cracked weld you can see. A failed test turns red. But a good decision and a bad decision look exactly alike on the day they are made — both are a confident sentence in a meeting, a line in the minutes, a nod around the table. The flaw does not surface for months, by which time the room scattered, the context is gone, and the person asking "who decided this?" meets a wall of shrugs. Decisions are the one product whose defects stay invisible at the moment of creation and turn expensive at the moment of discovery. You do not build a quality system for a problem you cannot see.

The second reason is more human, and we rarely say it out loud. Writing a decision down — who made it, why, and on what evidence — creates accountability, and accountability is uncomfortable. A documented rationale can be proven wrong in writing. A named decider can be held to the call later. So a quiet incentive threads through every organization to keep decisions a little vague, a little verbal, a little deniable. The blind spot is not only a gap we never tooled for; it is, in part, a gap we prefer. And that is precisely why a *system* must carry the discipline — because left to our own comfort, we will keep choosing the fog.

What the blind spot looks like from the inside

You do not need the statistics, though, because you live the symptoms. Let me show you four faces of the same gap. You will recognize every one.

The hall of doors. Picture a colleague — a project sponsor with a proposal she believes in. To get a yes, she steps into a process that feels, from the inside, like a hall of doors. First, a weekly meeting where business and technology size up everything seeking funding. Clear that, and as many as four more doors wait: an architecture review, a security review, an AI review if her project touches AI, and the Architecture Review Board itself if the change is big enough. Five gates. And here is the quiet cruelty: no one can tell her where she sits in the line, or when a decision will land. She enters the funnel and waits, with no map and no clock. That is not a uniquely broken company. That is a normal Tuesday at a large one.

The rubber stamp. Now picture the opposite failure, often in the same building. A change board convenes. Someone reads the title of a change. "Any objections?" Silence. "Approved." Next. A dozen of these in an hour, each one touching live systems and real money, and not one revisited later to ask the single question that teaches you anything: *was that the right call, and how do we know?* One process drowns the sponsor in gates; the other waves changes through without a glance. Opposite symptoms — and the same missing organ.

The needle in the field of haystacks. Six months after any decision, someone asks the reasonable questions: Who decided this? Why this option and not the others? Where is it written down? Were we sure? The honest answer is a slow hunt through a ticket title, a dead thread in a tool nobody opens, and a meeting whose notes never quite got written. The decision happened. The proof did not. Finding the rationale later is like finding one specific needle in a field of haystacks — so most of the time we give up and decide again from scratch, which is its own quiet tax, paid forever.

The argument we already lost. And here is the one that stings the most. A team debates a hard call for weeks, lands on an answer, and ships it — never knowing that two years earlier a different team faced the identical question, tried the identical answer, and watched it fail. The lesson existed. It simply lived nowhere anyone could reach. So the organization pays full price for the same mistake twice, three times, as many times as the people who remember keep leaving. A company with no decision record does not accumulate wisdom; it accumulates turnover. Every resignation is a small house fire in a library no one backed up.

But we already log decisions, don't we?

Now, if you run projects, you might be bristling a little — because you *do* keep a kind of decision record, and you are right to say so. Most projects maintain a RAID log: Risks, Assumptions, Issues, and Decisions. The Decisions tab is exactly what it sounds like — a running list of what the steering committee, or more often a lone project manager, chose to adopt. Tactical calls, strategic ones, architecturally significant ones, budget calls, customer-experience calls, all in one column. There is usually an owner. And if the template is unusually disciplined, it might even hold the options considered and the outcome expected — though, in practice, that is a rare and beautiful sight.

Agile keeps its own version. The retrospective is where a team writes down what worked, what did not, and what to change — lessons that, done well, loop back into the next sprint as fresh decisions.

So the instinct is already there. We try. And that makes the gap sharper, not softer, because of three questions a RAID log and a retro cannot answer.

Who keeps track of all of them? Each log lives inside its own project, and when the project ends, the log dies with it — archived in a folder no one opens, on a drive no one inherits.

Does anyone ever go back to read one to inform a new decision? Be truthful. Maybe once. And if they wanted the real context behind a one-line entry — the debate, the dissent, the evidence, the reason the obvious option lost — how far would they actually get before the trail went cold?

That is the problem in a sentence. We do not lack the instinct to record decisions. We lack a place where those records live past the project, connect across projects, keep their context, and can actually be asked a question. A RAID log is a decision diary that no one rereads. What the blind spot needs is a decision *ledger* that the whole enterprise keeps — and that is a different thing entirely.

The flood is rising

If the blind spot were holding still, we could get to it eventually. It is not holding still. It is growing, fast, and Artificial Intelligence (AI) is the reason.

In one company I know, a team switched on the meeting assistant in their video calls — the little agent that takes the notes. A marvel, at first: no one scribbles anymore, and at the end there is a tidy summary of what was discussed, what was decided, and who owns what. Then you do the arithmetic. Meetings like that run every hour, in half the departments, all day. Each one produces decisions. Multiply it out, and you reach a number of decisions per day that no human team could ever read — let alone trace, record, or stand behind.

And the new decisions are not only more numerous; they are harder to see into. When a person makes a call, you can at least ask them why. When a model makes or shapes the call — ranking the options, drafting the recommendation, scoring the risk — the "why" arrives as a fluent, confident paragraph that may or may not correspond to anything real. The flood is rising, and the water is getting murkier at the same time: more decisions, from more hands and more machines, each one a little harder to inspect than the last.

There is a clock on this, too. Regulators are beginning to ask for exactly the trail we never kept — who decided, on what basis, with what human oversight, and can you prove it. The blind spot is sliding from an efficiency problem into a compliance one. AI is not relieving the decision load. It is multiplying it, speeding it up, and clouding it — and aiming all three at the one part of the enterprise that never got a bucket.

The shape of it

Put the pieces together and the blind spot takes a clear shape:

We cannot improve what we do not measure.

We cannot defend what we did not record.

We cannot learn from what we never revisit.

Every other expensive thing in the building is measured, assured, and on the record. The decision — the most expensive thing of all — is none of those. We fly the most important instrument in the cockpit with the dial taped over, and then act surprised when we end up somewhere we never meant to go.

So what?

Here is what the blind spot costs *you*, specifically, in your chair.

You know the scene, because you sat in it. A decision from a year ago turned out badly, and the meeting is called. The people in the room are not the people who made the call; the people who made the call are gone, or do not remember, or remember three different versions. Someone pulls up a ticket with a one-line title. Someone else finds a thread that trails into nothing. The actual reasoning — the options weighed, the risk knowingly accepted, the quiet voice that said "I am not sure about this" — is simply not anywhere. So the meeting does the only thing it can: it guesses, it assigns blame by mood, and it resolves to "do better next time," with no mechanism on earth to make that true.

That is the tax. Every decision you cannot defend is a future meeting exactly like that one. Every rationale that lives only in memory walks out the door the day its owner does. Every mistake you fail to record, you quietly license yourself to repeat. The blind spot never announces itself; it bills you slowly — in rework, in repeated errors, in the slow erosion of trust between the people who decide and the people who live with the results. You are not paying for one grand catastrophe. You are paying a thousand small, invisible tolls, and you will keep paying them until you choose to see them.

What now?

One small move this week, and it costs you nothing but a column or two.

Start a decision ledger. Take a spreadsheet, a shared list, or a page in whatever tool you already use, and for every consequential decision you make, write down five things *as it happens*: what was decided, who decided it, what options were weighed, why this one won, and — the field everyone forgets — what outcome you expect, so you can check yourself later.

Do not boil the ocean. Start with the decisions that would hurt the most to get wrong — the five or ten a quarter that move real money or real risk. That is the entire assignment. You are not buying a platform; you are doing for your decisions exactly what a bookkeeper does for money: recording each one as it occurs, at the moment it occurs. Accounting calls that a general ledger. Yours is a *decision ledger* — and the day you start it is the day your blind spot begins to close. Nearly everything else in this book builds on that one humble habit.

Where we go next

Of course, staring at a flood this size, there is a far more tempting answer than a humble ledger — and every vendor alive is selling it. If decisions are too many and too fast for people to keep up with, why not hand them to a machine that never tires and never sleeps? Build a smarter board. Automate the gate. Let the AI decide.

It is a seductive idea. A friend of mine bet his job on it. The next chapter is the story of what that promise actually delivers.

Chapter 1 in review

Key takeaways

The decision is the **largest unmeasured surface** in the modern enterprise: we instrument the web page to the millisecond and leave the funding call in the dark.

The blast radius is enormous — roughly **$250 million a year** of wasted manager time for a single large firm (McKinsey), and trillions in failed transformations — but the deeper cost is **decision debt**: every unrecorded choice compounds silently.

- The blind spot persists for two reasons: a decision **hides its defects** (a good and a bad call look identical on the day), and recording one **creates accountability** people quietly prefer to avoid.
- It shows four faces: the **hall of doors** (gates with no map), the **rubber stamp** (verdicts no one revisits), the **needle in the haystack** (rationale you cannot find), and **the argument we already lost** (institutional amnesia — the same mistake, paid for twice).
- We *do* try to record decisions — **RAID logs** and **retrospectives** — but they die with the project, never aggregate, and rarely keep their context.
- AI is widening the gap fast: more decisions, faster, and more **opaque**; and regulators are beginning to demand the very trail we never kept.
- The first move costs almost nothing: start a **decision ledger** — record each consequential decision as it happens.

Key terms

Decision debt — the compounding cost of decisions made with no record, rationale, or way to learn from them; the quiet cousin of technical debt.

Decision ledger — a running record of decisions, captured as they occur, the way a bookkeeper records transactions in a general ledger.

RAID log — a project register of Risks, Assumptions, Issues, and Decisions.

Retrospective — the agile practice of reviewing what worked and what did not, to improve the next iteration.

Institutional amnesia — an organization's loss of its own hard-won lessons as the people who held them move on.

Check your understanding

Why does the author call decisions "the largest unmeasured surface" in the enterprise?

What is *decision debt*, and how is it like technical debt?

Give two reasons the blind spot persists.

Name the four "faces" of the blind spot.

Why are RAID logs and retrospectives not enough to close it?

Answer key: (1) Every other expensive output is measured and assured; the decision — the most expensive — is not. (2) The compounding cost of unrecorded, ungrounded decisions; like technical debt, it accrues silent interest and comes due later. (3) A decision hides its defects (good and bad look alike at the time); and recording creates accountability people avoid. (4) The hall of doors, the rubber stamp, the needle in the haystack, and the argument we already lost (institutional amnesia). (5) They are per-project, abandoned at project end, never aggregated, and thin on context.

Questions for reflection and discussion

Which of the four faces is most familiar in your organization, and what does it cost?

Where is your largest pocket of decision debt — a foundational choice no one can reconstruct?

For facilitators: ask the group whether adding *more gates* ever made their decisions sounder. Use the answer to separate "more control" from "more assurance."

Apply it

Start your decision ledger this week. Take a spreadsheet or shared page, and for the five-to-ten decisions that move real money or real risk this quarter, capture five fields *as each one happens*: what was decided, who decided, what options were weighed, why this one won, and the outcome you expect. *Trainers:* collect anonymized ledger rows from the cohort and discuss which field people skip most. It is usually the expected outcome — the one that makes learning possible.

Notes and further reading

McKinsey & Company, "Decision making in the age of urgency" *(≈530,000 manager-days / ≈$250M per year per large firm).*

The "$2.3 trillion" figure refers to **failed digital-transformation programs worldwide**, used here as a symptom, not a direct decision-cost.

The decision-ledger idea is developed further in the chapter on Memory.

CHAPTER 2

The Seduction of the Smarter Machine

A friend of mine runs the Architecture Review Board at a national retail brand. Good engineer, good leader, badly underwater. His board is so backed up that a new proposal waits three months just for a slot on the agenda — three months before anyone even looks at it. The complaints reach him daily. And his own job, he is told, depends on fixing it.

So one afternoon he gathers his team and gives a single, understandable order: automate everything. Use whatever AI you can get your hands on. Build me a review board that does what the human board does, only faster, so the wait disappears. Off they go, chasing a kind of nirvana — an artificial board that reviews, judges, and rules on architecture around the clock, no backlog, no bottleneck.

I understand the impulse completely. I felt it myself. And that is exactly why I want to slow it down — because the order my friend gave is the most natural, most tempting, and most dangerous move in this entire book.

The most reasonable bad idea in the building

Here is the seduction, stated plainly. You saw the setup in the last chapter: decisions are too many, too fast, too overwhelming for people to keep up with. So the obvious answer arrives, gift-wrapped, from every vendor and every conference stage — if people cannot keep up, take the people out of the loop. Let the machine decide.

And the case for it is strong. Let me make it as well as its champions would, because a seduction you dismiss too quickly is one you never really understood.

A machine never tires. It does not get decision fatigue at four in the afternoon. It does not drag yesterday's argument into today's call. It applies the same criteria to the thousandth case as to the first. It scales without hiring. It is fast — a verdict in seconds, not a slot in three months. And it can be made consistent in a way a rotating cast of humans, with their moods and their politics and their Tuesdays, never will be. Point all of that at a clogged review board and the promise is intoxicating: clear the backlog, cut the cost, free your best people from the drudgery of the gate.

There is history on the seduction's side, too, which is why it feels less like a gamble than an inevitability. This is the latest verse of a very old song. We automated the loom, and the weavers were right to worry, and the cloth got cheaper anyway. We automated the assembly line, the switchboard, the typing pool, the back office, the trading floor. Each time, the same fear; each time, the machine won, and mostly we came out better for it. So when someone says "automate the decisions too," they are not being reckless. They are pattern-matching to two centuries of automation that paid off. The burden of proof feels like it sits on whoever says *wait*.

So let me carry that burden, and name the one thing that makes a decision different from a bolt or a trade. A bolt carries no accountability. When a machine stamps a defective bolt, you scrap the bolt; nobody asks the bolt to justify itself. A decision is not like that. A decision carries answerability — a human or an organization on the hook for it, to a regulator, a board, a court, a customer. You can automate the *making* of a thing that answers to no one. You cannot, in the same way, automate the making of a thing that someone must answer *for*. That difference is the reason this old song needs a new verse.

If the story ended there, I would be selling you the machine, not warning you about it. It does not end there.

The forms the seduction takes

The temptation rarely shows up wearing a sign that says "let the robot decide." It arrives in respectable clothing, and three outfits in particular are everywhere right now.

The first is the **copilot** — an assistant that drafts the recommendation, ranks the options, scores the risk. Useful, and mostly honest about being an assistant. The human still signs.

The second is more advanced, and it is sweeping through the industry as I write this: the **LLM-as-judge** — a Large Language Model (LLM) pointed not at making the work but at *grading* it. Give it a rubric and it evaluates outputs, rates answers, approves or rejects, at a scale no human reviewer could touch. In the language we will lock down in a moment, the LLM-as-judge is *Decision Control, automated* — the inspector at the gate, replaced by a model. It is useful, and it is spreading fast.

The third is my friend's dream: the **autonomous board** — not one judge but a panel of agents, deliberating with each other, reaching a collective verdict, and rendering it with no human in the room. The whole review board, lifted out and rebuilt in software. The throughput is irresistible.

Each of these moves the same direction: take the slow, expensive, tired human out of the decision, and let the tireless machine carry more of it. And each runs into the same wall.

The consistency trap

Go back to the strongest line in the machine's favor — the one even skeptics concede. *It is consistent.* It applies the same standard every time, where a room full of tired humans never will. That is true, and it is exactly where the danger hides.

Consistency is a virtue only when the rule is right. When the rule is wrong, consistency is the thing that turns a small mistake into a catastrophe. A human board is gloriously inconsistent: one reviewer misses what another catches, today's panel runs harsher than last week's, and that very unevenness means a bad call often gets caught by the next set of eyes. A machine offers no next set of eyes. When a model's judgment is subtly off — a bias in its training, a blind spot in its rubric, a wrong assumption baked in — it does not make that mistake sometimes. It makes it *every single time*, with perfect uniformity, ten thousand times a day, and it never gets tired enough to slip and accidentally get one right. Consistency without assurance is not safety. It is industrialized bias, applied flawlessly, at scale.

The catch nobody prices in

Here is the question the seduction never answers, and it is the only one that matters: when the machine decides, and the machine is *wrong*, what then?

Because the machine will be wrong, sometimes, and in a particular way that should worry you more than ordinary error. A model does not fail like a tired human, with a yawn and a visible shrug. It fails *fluently*. It produces a wrong answer with the same confidence, the same polished paragraph of reasoning, as a right one. The industry word for the worst version of this is *hallucination* — a fabricated fact or a fabricated justification, delivered with total composure. And the trouble with a confident, fluent error is that it does not look like an error. It looks like an answer.

Consider the day a single wrong sentence cost a hundred billion dollars. Late in 2022, ChatGPT stunned the world, and Google — the company that defined search for a generation —

suddenly looked slow. So in February 2023, under enormous pressure to prove it still led, Google unveiled its own chatbot, Bard, to the public in a polished promotional advertisement. The ad was meant to charm: someone asks Bard what to tell a nine-year-old about the discoveries of the James Webb Space Telescope, and the chatbot answers warmly, fluently, helpfully — a little burst of wonder about the cosmos. And in the middle of that warmth, it states, plainly, that the telescope took "the very first pictures of a planet outside of our own solar system."

It is a lovely sentence. It is also false. The first image of an exoplanet belongs to 2004, to an entirely different instrument, years before Webb ever launched. But the false sentence carried no flag, no hesitation, no tell. It sat in the paragraph wearing the same calm confidence as the true sentences around it — which is the entire problem. Astronomers spotted the error within hours, on the very day of a campaign meant to reassure the world that Google still owned this technology. Reuters reported it. And the next day, Alphabet's stock closed down almost eight percent — erasing something on the order of a hundred billion dollars in market value, over a single wrong clause in an advertisement the company itself wrote, reviewed, and approved.

Sit with that proportion. Not a defective product. Not a data breach. Not a scandal of conduct. One fluent, confident, unverified sentence, released at the scale of a global launch — and the markets answered with the wealth of a small nation. The error never announced itself; it looked exactly like the dozens of correct facts a chatbot produces every minute. And so no one in the entire chain — not the engineers, not the marketers, not the executives who signed off on the most scrutinized advertisement of their year — asked the one question that would catch it: *what is the source for this claim, and does it hold up?* They never asked, because the sentence did not look like it needed asking. That is precisely what fluent failure does: it disarms the very suspicion that would catch it. One grounding step stops that clause cold — *cite the source, or withhold the claim* — before it reaches the ad, let alone the market. The cost of running the check was a few seconds. The cost of skipping it ran to eleven figures.

Now scale that. A human board that rubber-stamps a dozen decisions an hour is a problem. A machine board that confidently rules on ten thousand — a few of them fluently and invisibly wrong — is a different category of problem: faster, quieter, far harder to catch. Automating the decision does not remove the risk of a bad decision. It industrializes it.

And the LLM-as-judge, the automated inspector, does not escape this — it inherits it. A judge is simply another decision-maker, and it can be confidently wrong about its verdicts too. Which raises a small, devastating question that the whole "let the model grade it" movement tends to skip: *who judges the judge?*

And do not reach for the obvious fix, because it does not bottom out. If the worry is that the judge might be wrong, add a second model to check the first. Now what checks the second? A third. It is checkers all the way down, and the regress never terminates on its own. It stops only when you reach something that cannot bluff — a deterministic rule that simply *is* what it is — or someone who can be held accountable and is therefore motivated to be right. Stacking models does not create trust. It defers it. Trust must land somewhere a model cannot, and that somewhere is the whole subject of this book.

The oldest irony in the book

None of this is new, by the way. Engineers learned it forty years ago, in control rooms, long before anyone said "AI."

In 1983, a researcher named Lisanne Bainbridge wrote a short, devastating paper called "Ironies of Automation." Her observation was simple, and it aged frighteningly well. When you automate most of a process but leave a human responsible for the rest, you do not make that human's job easier. You make it harder. The operator now spends the day monitoring a system that runs itself — dull, until the rare moment it fails, at which point they must suddenly understand, diagnose, and correct a process they were not actively running and may no longer fully grasp. The more you automate, the more critical and the more demanding the human's leftover role becomes, not less. And — her sharpest point — the human can only step in if the system stays *transparent* enough to be understood.

Read that across forty years and it lands squarely on us. Automate the deciding, and you do not retire the human. You promote them — from making the calls to assuring them, from in the loop to above it — and you hand them a harder job: catching the rare, fluent, automated mistake inside a flood of automated decisions. That job is impossible without exactly the thing the seduction forgets to build: transparency, evidence, a record — assurance. Bainbridge saw the shape of our problem in a power plant in 1983. We just bolted a much bigger engine onto it.

The morning the bill comes due

Let me make it concrete, because the abstract version is too easy to wave away.

Imagine a company — pick any fast-moving, AI-first one you like — that did the seductive thing and let its systems make a class of consequential calls automatically, at speed, with no human in the chair and no record of *why* each call was made. For a long while it works beautifully. Throughput soars. Costs fall. Everyone is happy.

Then one morning a regulator, or a court, or a furious customer, or a government official points at one of those automated decisions and asks the simple question: *why did you do that, and can you prove it was sound?* The company opens its hands and finds nothing. No rationale. No record. No human who made the call and can defend it. Just a log that says the system decided, on a date, at a time. "The AI decided" is not a defense. It was never a defense. Somebody — a human, an executive, a board — absorbs the blow, and learns in the worst possible moment that you cannot delegate *accountability* to a machine, no matter how much of the *deciding* you handed it.

And notice the cruel asymmetry of speed. The machine produces decisions faster than any human can review them, which means the unreviewed, undefended liability does not merely sit there — it *compounds*, decision by decision, hour by hour, while everyone admires the throughput. By the time the question arrives, the pile of things you cannot defend is not a stack of papers. It is a mountain you did not know you were building.

That is the bill the seduction never shows you on the way in. The machine can take the decision. It cannot take the *answerability* for the decision. That stays with a person, always — and the faster and more autonomously the machine decides, the more of it stacks up on that person, unrecorded and undefended, waiting for the morning someone asks why.

And lest that scene feel hypothetical, a Canadian tribunal already lived it — and the human at the center of it was simply grieving. After a loss in his family, a man named Jake Moffatt went to Air Canada's website late one night to book a flight to the funeral, and, unsure how the airline's bereavement fares worked, he asked the site's chatbot. The bot answered with perfect, immediate confidence: go ahead and book now, it told him, and you can apply for the bereavement discount retroactively, within ninety days of the flight. It even linked him to the airline's own bereavement-policy page — a page that, read plainly, said the exact opposite. Moffatt did the reasonable thing. He trusted the answer the airline's own website gave him, booked the full fare, and later applied for the refund the bot promised.

Air Canada refused. The chatbot, it explained, was simply mistaken — and then the airline reached for a defense that deserves to be taught in business schools for a generation. It argued, in front of a tribunal, that the chatbot was "a separate legal entity that is responsible for its own actions." Read that again. A company built a tool, put it on its own website, let it speak to its own customers in its own name — and then, when the tool was wrong, told a court the tool was on its own. The machine, the airline suggested, should answer for itself.

The tribunal did not blink. It rejected the argument outright as something close to absurd, found that Air Canada committed negligent misrepresentation, and held the airline responsible for every word that appeared on its own website, whether a human or a chatbot produced it. Moffatt won his refund — about eight hundred Canadian dollars, a rounding error to an airline — and the case became the precedent every technology lawyer now cites. Hold the size of the award against the size of the principle: the money was trivial, the ruling enormous. A court stated, in law, the exact rule this book states in principle — a machine's output is the organization's decision, and accountability does not transfer to the machine that produced it. *"The AI said so"* is not a defense, not in a courtroom and not in your next audit. And notice the failure underneath the law: the bot's confident claim contradicted the company's own written policy, sitting one click away. The ground truth was right there, on the same website, the whole time. Nothing in the system made the machine check itself against it.

The law is writing it down

You might read that last scene as a worst case — vivid, but rare. It is fast becoming the baseline. The regulator in that story is no longer hypothetical; lawmakers are turning "why did you decide that, and can you prove it?" from an awkward question into a legal requirement.

Europe's General Data Protection Regulation already gives a person who is subjected to a significant, solely automated decision the right to human intervention, to express their view, and to contest it. Read the fine print and it gets sharper: a decision still counts as "solely automated" even when a human signs off — if that human merely rubber-stamps the machine's output without a real, independent assessment. In other words, the rubber-stamp board from the last chapter is not a human-in-the-loop safeguard in the eyes of the law. It is an automated decision wearing a person as a costume, and it carries every obligation that comes with one.

The European Union's AI Act goes further for "high-risk" systems — the kind that rule on credit, employment, and essential services. It requires meaningful human oversight, transparency into how the system reaches its conclusions, and automatic record-keeping: logs, kept for a defined period, that let a decision be traced and examined after the fact. Other jurisdictions are moving the same direction, toward one demanding principle — if an automated system makes a consequential decision about a person, a human must be able to step in, the logic must be explainable, and the trail must exist.

The law is writing it down because it already watched the alternative, twice, in slow motion. Two cases sit behind these new statutes like cautionary monuments — and note carefully what they are, because the distinction protects the whole argument: these are not chatbot hallucinations but their older, colder cousin, automated systems deciding at scale, on rigid rules instead of evidence, with no human standing above the loop.

In the Netherlands, the tax authority ran a self-learning algorithm to hunt for fraud in childcare benefits. Quietly, it learned to treat ordinary facts about a family — a second nationality, a low income — as marks of suspicion, and it began flagging families of Moroccan, Turkish, and Surinamese descent in their thousands. These were not abstract flags. Roughly twenty-six thousand families were branded fraudsters by the system and ordered to repay benefits going back years — sums that emptied savings, forced foreclosures, broke marriages. More than sixteen hundred children were taken from their homes into state care while their parents fought an accusation they could not even see, because the basis of it was a model no official could fully explain and no family could question. The damage ran for years before anyone in power admitted what the machine did. When the reckoning finally came, the Dutch data-protection authority ruled the entire apparatus unlawful and discriminatory, and in January 2021 the whole government of the Netherlands resigned over it. An algorithm, ungrounded and unaccountable, brought down a national government.

Across the ocean, Michigan ran a system with the bland name MiDAS to judge unemployment-fraud cases — and it judged them with no human reviewer in the loop at all. Any mismatch in the data, any clerical discrepancy, was read automatically as fraud, and the machine moved straight to punishment. The state auditor later pulled a sample of its findings and discovered that ninety-three of every hundred were wrong. Ninety-three percent — a system wrong far more often than right, running unsupervised on people's livelihoods. By the time anyone caught it, the machine falsely accused some forty thousand people of fraud; the state seized their tax refunds, garnished their wages, and stacked penalties as high as four times the amount supposedly owed — all before a single human being ever looked at a single case. The class action that followed eventually settled for twenty million dollars. These are precisely the decisions the new laws exist to prevent: consequential, automated, aimed squarely at a person, and impossible to contest because no one kept a record a human could read and challenge. In both, the machine decided, at scale, and no one assured it — which is exactly the gap (grounding, the record, and a human kept above the loop) that the statutes now convert from good practice into legal duty.

Now sit with what that means for the seduction. The very things it quietly discards — the human who actually engages, the rationale, the record — are precisely the things the law is starting to mandate. Automate the deciding without assurance, and you are not just building a liability. You are building one with a statutory deadline. The audit trail you skipped in order to go faster becomes the exact document a regulator will one day require you to produce — and you will not be able to.

What the seduction gets wrong

Step back and you can see the deep mistake. The seduction assumes the bottleneck is the *deciding* — that the problem is simply that humans are too slow at rendering verdicts, and the cure is a faster verdict machine. So it automates the verdict.

But re-read my friend's actual problem. His pain is not only the speed of the verdicts. It is the volume to be tracked, the absence of any record, the inability to learn from last quarter, the dread of the audit, the proposal that sits for three months with no map and no clock. Almost none of that is fixed by deciding faster. A machine that renders verdicts at ten thousand an hour, with no assurance underneath, hands him a faster, larger, more opaque version of the exact problem he started with. He does not need a quicker gate. He needs the thing the gate always lacked: assurance — a way to make every decision, his or the machine's, complete, consistent, governed, evidenced, and on the record, with a human still answerable for it.

That is the turn this whole book makes. Not "should the machine decide?" but "whoever decides, can we *assure* it?" Point the AI at the assurance — the tracking, the checking, the evidence, the record, the memory — and you relieve the real bottleneck without ever handing away the accountability. Point it at the deciding alone, and you build a faster way to be undefendable.

So what?

For you, in your chair, the seduction is not a far-off philosophical risk. It is a proposal that will land on your desk this year, maybe this quarter, wrapped in a compelling demo and a tempting price — and you will be under pressure, like my friend, with your name on the delay, to say yes. What is at stake in that moment is your judgment: the difference between buying a machine that decides *for* you, and buying one that *assures* the decisions you remain accountable for. One clears your backlog and quietly mortgages your defensibility. The other clears your backlog and strengthens it. They can look identical in the demo. They are opposites in the audit.

What now?

When the smarter machine shows up — and it will — ask it three questions before you fall in love.

One: *when you are wrong, how will I know?* A tool that cannot tell you where it is uncertain, or show you the evidence behind its call, is a tool that hides its errors inside its confidence.

Two: *who is accountable for what you decide?* If the honest answer is "the system," walk away. The answer you want is "you are — and here is the trail to prove you decided well."

Three: *do you make my decisions, or do you assure them?* Write the answer down. It is the difference between Decision Control handed to a black box and Decision Assurance handed to you.

Where we go next

Now — is the seduction actually wrong, or am I just the cautious type, telling you to slow down while the world speeds up? Fair challenge. You should not take my word for it, and you need not, because I did not take my own word for it either. I believed a version of the dream — that a well-built board of AI agents might really out-decide the expert in the chair — enough to stop arguing about it and go find out. I built the machine. I ran the experiment the rigorous way, with the hypothesis written down in advance so I could not move the goalposts afterward. The next chapter is the story of what happened when I tested the seduction on myself — and what the data did to my favorite idea.

Chapter 2 in review

Key takeaways

- The seduction is simple: decisions overwhelm us, so **let the machine decide**. Every vendor sells it, and it feels like the next verse of a two-century automation success story.
- But a decision is not a bolt. A bolt carries no **accountability**; a decision always does — and you cannot automate the making of a thing someone must answer *for*.
- The seduction wears three outfits: the **copilot**, the **LLM-as-judge** (an automated inspector — Decision *Control,* automated), and the **autonomous board**.
- The machine's prized **consistency** cuts both ways: consistency without assurance is industrialized bias, applied flawlessly at scale.
- Machines fail **fluently** — a confident, fabricated answer that looks like an answer (**hallucination**). Automating decisions does not remove the risk; it industrializes it. And *who judges the judge?* — the regress stops only at something deterministic or someone accountable.

- The **irony of automation** (Bainbridge, 1983): automating most of a job makes the human's residual role *harder*, more critical, and dependent on transparency.
- The law now demands the trail (GDPR's right to human intervention; the EU AI Act's oversight, transparency, and logging duties). The fix is not a faster decider; it is **assurance** — point the AI at assuring, not at deciding.

Key terms

Hallucination — a confident, fluent, fabricated output (a fact or a rationale) presented as if true.

LLM-as-judge — using a large language model to grade or rule on outputs; in this framework, Decision Control automated.

The irony of automation — automating most of a process leaves the human a harder, more critical, monitoring-and-rescue role (Bainbridge, 1983).

Accountability vacuum — the gap that opens when an autonomous system decides and no human can answer for the call.

GDPR Article 22 / EU AI Act (high-risk) — laws requiring human involvement, explainability, and record-keeping for consequential automated decisions.

Check your understanding

- Why is automating a *decision* different from automating a *bolt*?
- Name the three forms the seduction takes, and say which one is "Decision Control, automated."
- What does it mean that a machine "fails fluently," and why is that dangerous?
- State the irony of automation in one sentence.

Under GDPR, does a human rubber-stamping an AI output count as meaningful human involvement?

Answer key: (1) A bolt carries no accountability; a decision does, and answerability cannot be automated away. (2) Copilot, LLM-as-judge, autonomous board; the LLM-as-judge is Control automated. (3) It produces wrong answers with the same confidence as right ones, so the error looks like an answer. (4) Automating most of a job makes the remaining human role harder and more critical, not easier. (5) No — a rubber-stamp still counts as a solely automated decision.

Questions for reflection and discussion

Where did you watch automation move a bottleneck downstream rather than remove it?

The chapter ends with three questions to ask any decision tool. Which would most embarrass the tools you use today?

For facilitators: make the strongest possible case *for* letting the machine decide, then ask the room to find the missing word. (It is *accountability*.)

Apply it

Take a real or proposed AI decision tool and put it through the chapter's three questions: *When you are wrong, how will I know? Who is accountable for what you decide? Do you make my decisions, or do you assure them?* Write down each answer. A tool that cannot answer the first two, or that answers "the system" to accountability, is a liability with a demo. *Trainers:* run this against a live vendor sheet and score it together.

Notes and further reading

Bainbridge, L. (1983). "Ironies of Automation." *Automatica* 19(6), 775–779.

GDPR Article 22 (automated individual decision-making); EU AI Act, high-risk obligations (human oversight, transparency, logging).

On LLM-as-judge reliability (position, verbosity, self-preference biases), see Zheng et al. (2023) — to cite at polish.

Woven cases: Google Bard's James Webb error (Feb 2023; Alphabet shares −7.7%, ≈$100B) — Reuters; CNN. *Moffatt v. Air Canada* (BC Civil Resolution Tribunal, 2024; ≈CAD $812; the "separate legal entity" argument rejected) — 2024 BCCRT 149 (CanLII). The Dutch childcare-benefits scandal (*toeslagenaffaire*; ≈26,000 families, >1,600 children taken into care, government resigned Jan 2021). Michigan MiDAS (state-auditor 93% error rate, ≈40,000 wrongly accused, $20M settlement). **Note:** Bard and Air Canada are generative (LLM) hallucinations; the Dutch and Michigan cases are *algorithmic* automated-decision failures — a distinct, pre-LLM family, labeled as such to keep the argument honest.

CHAPTER 3

What Happened When I Tested It on Myself

I owe you a confession, and it is the reason you can trust the rest of this book.

I wanted the smarter machine to win.

For twenty-five years I studied complex systems — the way an ant colony, with no genius ant in charge, solves problems no single ant could; the way a market, full of fallible traders, prices the world; the way a flock of birds turns as one. I built a whole intellectual life on a single, beautiful idea: that a group, organized well, becomes more than the sum of its parts. The crowd is wise. Emergence is real. So when the seduction of the last chapter arrived at my own door — *let a board of AI agents make the decisions* — I did not recoil from it. I leaned toward it. I suspected it was true. A panel of AI agents, each a little different, deliberating together, *should* out-decide any one of them alone. That is what groups do. I expected the machine board to be wiser than its smartest member, and I wanted to be the one who proved it.

That conviction was not a hunch; it was a library. Horst Rittel and Melvin Webber, back in 1973, named the "wicked problems" that defeat ordinary analysis — and for fifty years the field's answer ran the same way: you meet a wicked problem with many minds, not one. Scott Page proved, with actual mathematics, that a diverse group can out-predict its best individual member — the "diversity trumps ability" result that launched a thousand keynotes.

James Surowiecki gave it a household name: the wisdom of crowds. I cited this work in my own papers. I built frameworks on it. I stood on conference stages and argued that the future of hard decisions was collective, emergent, distributed. So you can see why I did not approach the AI board as a skeptic. I approached it as a believer looking for his proof.

So I did the one thing that makes wanting something safe. I tested it in a way that would not let me cheat.

How to keep from fooling yourself

Here is the trap with a beloved idea: if you go looking for evidence that it is true, you will find it. You will run the test, get a messy result, and your mind — helpfully, and against your own interest — will reach for the version of the analysis that flatters the idea you already love. Scientists know this temptation intimately, and they invented a discipline to defend against it. It is called *pre-registration*, and it is simpler than it sounds.

Before you run the experiment, you write down — in public, with a timestamp you cannot alter — exactly what you are testing, exactly how you will measure it, and exactly what result would prove you *wrong*. Then you run it. The timestamp is the whole point. It means you cannot move the goalposts after you see where the ball lands. You cannot quietly redefine success to match the outcome. You committed, in advance, in ink, to letting reality get the final word.

I pre-registered my hypothesis on a public scientific registry, with a permanent identifier, before I ran a thing. I wrote down the claim — a board of AI agents decides better than its single best agent — and I wrote down, in advance, exactly what would count as that claim failing. Then I built the board and let it run.

The experiment, in plain terms

The setup was about as fair to my hypothesis as I could make it. I took real review decisions — the kind of accept-or-reject calls that expert panels make on technical proposals and on submitted research — and I ran them two ways. Once through a *board* of AI agents, several models reasoning together and reaching a collective verdict. And once through the *single best* agent on its own. Then I scored both against the verdict the human experts actually reached, and I asked the only question that mattered: did the board beat its best member?

I did not run this once. I ran it seven different ways, because I did not want a single lucky or unlucky result to settle it. I varied the models. I mixed different AI families together so the agents would think differently from one another — diversity is supposed to be the engine of collective wisdom, so I engineered as much of it as I could. I added a devil's advocate to the board, an agent whose only job was to argue against the group. I tried, in good faith and repeatedly, to give the collective every advantage the theory says it should need.

Let me show you one of the seven

The summaries are easy to nod along to, so let me make one of the seven concrete.

In one regime I used academic peer review — real papers submitted to a real conference, each one already carrying the verdict the human program committee reached: accept or reject. I fed those same papers to my board of AI agents, each agent playing a reviewer with a different lens — novelty, technical soundness, clarity, significance — and let them deliberate to a collective decision. Then I ran the identical papers through the single strongest agent alone. Both got scored the same way: how often did each match the humans' actual call?

The single agent did well. The board did worse. And the *way* it did worse is the tell. On the borderline papers — the ones where reasonable reviewers disagree — the board kept finding reasons to reject. One agent worried about the evaluation; another about a missing baseline; a third about the writing. Each concern was fair on its own. Pooled together, they buried papers that the human committee, and the single agent, judged worth accepting. The board was not stupider than its members. It was more *anxious* than its members — and anxiety, aggregated, reads as rejection.

What the data did to my favorite idea

It did not work.

Across all seven setups — different models, different domains, deliberately diversified agents, devil's advocate and all — the board did not out-decide its best single member. Not once, in a way that held up. Sometimes the board merely tied its best agent. Sometimes, and this is the part that stung, it did *worse* — a panel of capable models reaching a poorer verdict than the strongest one of them would reach alone.

I want to be honest about how that felt, because the feeling is part of the lesson. Twenty-five years of conviction rode on the opposite result. And the data, which I publicly forbade myself from arguing with, said no. There is a particular loneliness in arguing a case for years and then being the person who publishes the evidence against it. I felt it. And underneath the sting, slowly, something better: the quiet that comes when you finally *know* a thing, rather than merely wish it.

Why the board lost

Once the disappointment cleared, the mechanism turned out to be fascinating, and it matters for everything that follows.

A single, well-calibrated agent is a careful judge. It approves what deserves approval and rejects what deserves rejection at roughly the right rate. Now assemble a board of them and pool their concerns — the natural, sensible thing a board does, gathering every objection in the room. What happens is that the objections *accumulate*. Each agent flags something; the board, dutifully, honors all of it; and the collective starts rejecting things a calibrated individual would pass. The group does not average toward wisdom. It piles up caution until its judgment tips.

I did not accept that on the first try. If the board lost because its members were too alike, then making them more different should rescue it — so I rebuilt the panel from different AI families, models with different training and different temperaments, and ran it again. No rescue. If the problem was the simple pile-up of objections, then a smarter voting rule — requiring two agents to agree before a concern could sink a decision — should rescue it. It made things worse, by silencing the lone agent who was sometimes right. If the board needed sharper challenge, the devil's advocate should help. It changed nothing. Four different rescues, aimed at four different theories of the failure, and the board never beat its best member in any of them. After the fourth, I stopped hunting for the rescue and started believing the result.

The board's problem, in a phrase, was *miscalibration under aggregation*. The very act of combining many good judgments broke the calibration each of them held on its own. Pool a group carelessly and you do not get the best of everyone. You get the sum of everyone's nerves.

There was one more discovery in the wreckage, and it pointed straight at the way out. When the agents erred, they often erred *together* — confidently, in chorus, fabricating the same kind of unfounded objection. More agents did not cancel the error; they amplified it. But when I forced each agent to *cite* the evidence for every claim it made — to point at the actual text and quote it, or stay silent — the fabrications dropped by roughly half. Read that twice, because it is the seed of the whole second half of this book:

the fix for a confident machine error was not more machines. It was grounding. Verification. Making the claim show its receipts. The board could not vote its way to the truth, but it could be made to *prove its way* toward it.

What this does, and does not, prove

Now I must do the thing I promised at the start and draw the boundary plainly, because a result is worth only what its evidence supports.

This experiment set machine against machine, never machine against human. Every test compared a *collective* of AI agents to the *best single* AI agent. The "best expert" the board kept failing to beat is a model, not a person. People entered the experiment only as the standard — the human panel's actual verdict was the yardstick both the board and the solo agent were measured against. So nothing here says a machine out-judges a person, and nothing here says a person out-judges a machine. That contest was never staged. What I tested, and all I tested, is whether *assembling* AI agents into a board makes them decide better than the best one alone. The answer, across seven tries, was no.

That narrowness is not a weakness in the finding. It is what makes the finding trustworthy. I claimed one specific thing, I pre-committed to what would refute it, and reality refuted it. A disconfirmed prediction that you wrote down in advance is the most honest result a person can produce — precisely because you could not fake it into agreeing with you.

The failure that built this book

Here is the turn, and it is the hinge of everything.

A disconfirmed hypothesis felt, for about a day, like a dead end. It was the opposite. It was a doorway.

Because the result did not say *AI is useless for decisions.* It said something far more precise and far more useful: *do not deploy the machine to out-decide your experts, because it will not — but that was never where its value was.* The board could not beat its best member at the *verdict.* Along the way, though, it did something else flawlessly: it tracked every concern, surfaced every objection, applied every rule the same way every time, and produced a complete, inspectable record of how the decision got made. It was a poor oracle and a magnificent *assurance* system.

That is the whole pivot of this book, and I did not arrive at it by preference. I was dragged to it by my own disconfirmed experiment. Stop asking the machine to be a wiser decider. Ask it to be a tireless *assurer* — to make sure that whoever decides, human or machine, does so completely, consistently, on the evidence, and on the record. The dream of the smarter machine died in my own data. What rose in its place was Decision Assurance.

And one more thing, for the believers like me. The experiment did not kill emergence; it disciplined it. The wisdom of crowds is real — but it is not automatic, and it is not free. It shows up only under conditions this kind of decision did not meet: independent errors, room for a lone right voice to prevail, a way to aggregate that corrects rather than accumulates. Strip those away and the crowd does not get wiser; it gets more nervous. That is not a refutation of emergence. It is a map of its edges — and knowing where a beloved idea stops being true is its own kind of progress.

And if you want to see, in miniature, exactly why the human must stay above the loop while the machine assures, I can show you — because it happened while I was writing this very book.

Sidebar — A True Story from the Making of This Book

I want to tell you something that happened while I was writing this very book. It put a chill down my spine, and then it put a smile on my face — and by the end you will see why I left it in.

I was drafting the part where I describe the experiment at the heart of this book. A sentence came back to me, clean and confident: that I built a system to test whether a board of AI agents could out-decide *a careful human expert.*

It read beautifully. It was also false.

We never ran that test. We never pitted the machine against a person. The real experiment asked a narrower, humbler question: could a board of AI agents out-decide its own best single, well-calibrated member — the strongest agent on its own panel? People entered only as the measuring stick, the standard the machines were graded against. "Versus a human expert" was a story the sentence told because that is the story everyone tells. It sounded most certain exactly where it was wrong.

Here is the part that matters. I caught it. Not because I am clever, but because I remembered a simple rule we wrote down at the very beginning — the standard that named, in plain ink, what the experiment actually compared. My memory raised its hand. The written record let me *prove* the point instead of merely sensing it.

Sit with the irony for a second. A book about keeping machine hallucinations out of your decisions nearly went to print carrying one of its own.

And that near miss is the whole argument of this book, performed live on its author. Two forces caught the error, and neither would manage alone. My judgment noticed something looked wrong — that is *control.* The standard we wrote down made the wrongness provable — that is *assurance.* Strip out the written standard, and my memory is only a feeling someone could talk me out of. Strip out the watchful reader, and the standard sits in a file nobody opens. Together, they were airtight.

Now the uncomfortable truth, because I promised you those. The catch depended on me happening to remember. Most hallucinations meet no such reader. So please do not draw the cozy lesson that humans will always catch the machine. We will not. The real lesson is the harder one: build the assurance so the catch does not depend on a hero with a good memory on a good day. Make the standard surface itself. Make the claim carry its own proof. That is the difference between getting lucky and being safe — and it is what the rest of this book is about.

I left the story in on purpose. It is the most honest thing I can show you: not a polished promise that the machine behaves, but a true account of the day it did not — and the simple discipline that saved us both.

And it happens to the most careful people alive

I caught my own machine's lie because I remembered a standard. Please do not take the comfortable lesson from that — that a sharp human will always catch the machine. The opposite is true, and the proof is humbling: the most credentialed, most careful, most expert people in the world walk into the same fluent trap, because care is not a verification step and a reputation cannot read a citation back to its source.

Consider Michael Cohen — a lawyer who spent a career in the highest-stakes rooms in America. In 2023, working to end his supervised release, he needed case law, and he turned to Google's Bard chatbot to find it. He thought of it, he later explained, as "a super-charged search engine" — a faster way to do what he already knew how to do. Bard handed him three perfect citations: real-sounding names, real-sounding courts, exactly the precedents his argument needed. He passed them to his attorney, who folded them into a federal motion and filed it.

Then a judge sat down to read them. "There was only one problem," Judge Jesse Furman wrote in the order that followed. "The cases do not exist." Picture that quiet moment: a federal judge, motion in hand, going to look up the controlling precedent and finding nothing there — no case, no court, no words — only a confident citation pointing into empty air. Cohen was not careless in the ordinary sense; he simply trusted a machine that sounds exactly as certain when it invents as when it recalls. The judge, finding negligence rather than bad faith, declined to punish him. The lesson sits in the near-miss: a famous, sophisticated lawyer, and the fabrication still sailed through to a federal filing — because nothing in his process made the machine prove its citations before they became his.

If that case is a warning, the next one is almost too perfect to be true. Jeff Hancock is a Stanford professor and one of the world's foremost experts on misinformation — on exactly how people are deceived, and on the dangers of artificial intelligence in particular. The state of Minnesota retained him to defend a new law against AI-generated election deepfakes, and he prepared an expert declaration, signed under penalty of perjury, making the case for why AI deception is dangerous. To help draft it, he used GPT-4o. And the declaration he swore to went out carrying citations to academic studies that do not exist — invented sources, in a sworn filing, from a world expert on invented content.

The other side checked the footnotes. The fabrications surfaced. And Judge Laura Provinzino, excluding his testimony, named the irony so precisely that it belongs in this book: a leading expert on the dangers of AI, undone by leaning too heavily on it — "in a case," she wrote, "that revolves around the dangers of AI, no less." The man hired to warn the court about machine deception was himself deceived by a machine, and signed his name and his oath beneath it.

Sit with what these two cases, and the one from my own pages, are all saying together. A criminal-defense veteran. A world authority on AI misinformation. A team writing a book *about* keeping machine hallucinations out of decisions. Three of the most relevant, careful, qualified people you could pick for the job — and all three nearly shipped a machine's fabrication into something that mattered. Expertise did not save them. Care did not save them. The only thing that ever saves anyone is the dull, mechanical step none of them ran: take the claim, find its source, confirm it resolves to something real — or strike it. Grounding is not a quality of the person holding the pen. It is a process you run on every claim, precisely because the person — however brilliant — cannot be trusted to catch what reads exactly like the truth.

So what?

For you, this chapter is permission — permission to stop chasing a promise that does not pay, and to redirect the same technology at a problem it actually solves. The next vendor who tells you their agents will *decide better* than your experts is selling you the idea my data killed. The honest, valuable version of the pitch is the quieter one: not "our machine decides better," but "our machine makes every decision provable." A tool that cannot tell the difference between those two claims does not understand what it is for.

What now?

Do one thing the next time someone — a vendor, a board, your own enthusiasm — proposes letting AI decide. Ask for the evidence that it decides *better*, and ask it the way I needed to learn to ask: what was the prediction, written down in advance, and what would prove it wrong? If the answer is a glossy demo and a confident story, you are looking at a belief, not a result. Treat it the way I learned to treat mine: as a thing that owes you proof.

Where we go next

So we are left with a discipline, not an oracle — assurance, not super-intelligence. Which raises the obvious question: what *is* assurance, exactly, and how is it different from the gatekeeping we already do? That distinction — between assuring a decision and controlling it — is the quiet idea that organizes everything else in this book. It is the subject of the next chapter, and once you see it, you will not be able to unsee it.

Chapter 3 in review

Key takeaways

The author *wanted* the AI board to win — a career built on complex systems and the wisdom of crowds — so he tested the belief the one way that does not let you cheat: **pre-registration**.

Pre-registration means writing down, in public and in advance, the hypothesis, the measure, and what result would prove you *wrong* — so you cannot move the goalposts after seeing the data.

The experiment compared a **board of AI agents** to its **best single agent** on real review decisions, scored against the human verdict, across **seven** different setups.

Across all seven, the board did **not** out-decide its best member; sometimes it did worse. The author's favorite idea did not survive.

The mechanism is **miscalibration under aggregation**: pooling many agents' concerns piles up caution until a well-calibrated judgment tips into over-rejection. Four different rescues all failed.

The boundary matters: the test set **machine against machine, never machine against human**. The "best expert" was a model. The work makes no claim that machines beat — or lose to — people.

The disconfirmation built the book: a board is a poor *oracle* but a magnificent *assurer*. The fix for a confident machine error was not more machines; it was **grounding** — making each claim cite its evidence.

Key terms

Pre-registration — committing a hypothesis, method, and falsification criteria to a public, timestamped record before running the study.

Super-additivity (the assembly bonus) — the claim that a group decides better than its best individual member.

Miscalibration under aggregation — the failure mode in which pooling many agents' concerns degrades a calibration each held alone, here producing over-rejection.

Grounding — requiring a claim to cite checkable evidence (or abstain), which roughly halved the agents' fabrications.

The boundary of the claim — the explicit limit that the experiment compared models to models, never models to humans.

Check your understanding

What is pre-registration, and what problem does it solve?

What two things did the experiment compare, and against what standard?

What was the result, and what is the mechanism behind it?

Did the experiment test whether machines decide better than humans?

What single intervention roughly halved the agents' fabrications?

Answer key: (1) Committing the hypothesis, measure, and falsification criteria publicly and in advance; it stops you from fooling yourself after seeing the data. (2) A board of AI agents versus the best single agent, scored against the human verdict. (3) The board did not beat its best member (sometimes worse); the mechanism is miscalibration under aggregation. (4) No — machine against machine only; humans were the standard, never a competitor. (5) Grounding — forcing each claim to cite its evidence or stay silent.

Questions for reflection and discussion

Recall a belief you held strongly that the evidence later overturned. What let you finally accept it — or what kept you from it?

Where in your organization do people "result" — judge a decision by its outcome rather than its process?

For facilitators: ask why a *disconfirmed* hypothesis can be a stronger result than a confirmed one. (Because you could not fake it.)

Apply it

Borrow the pre-registration discipline for one real decision or forecast this month. Before you act, write down — where a colleague can see it — what you expect, and what observable result would tell you the call was *wrong*. Revisit it when the outcome arrives. *Trainers:* ask the cohort to pre-register a shared prediction at the start of the course and open the envelope at the end.

Notes and further reading

The pre-registered study lives on the Open Science Framework (OSF), DOI 10.17605/OSF.IO/ZWM3S.

On the wisdom of crowds and its conditions: Surowiecki (2004); Page's "diversity trumps ability" result; Rittel and Webber (1973) on wicked problems.

The full boundary statement: *DecisionAssurance_Ch3_Boundary_ThreatsToValidity_2026-06-21*.

PART II

The Paradigm Shift

CHAPTER 4

Assurance vs. Control

We ended the last chapter with a humbler machine than the one I went looking for — not an oracle that decides better, but an assurer that makes deciding *provable*. That word, *assurance*, is about to do an enormous amount of work in this book, so let me make sure it is not just a word. Let me take you back to the factory floor, because they drew the exact distinction we need a century ago, and they drew it in two people.

Walk onto any floor that makes something you would trust your life to — an aircraft brake, an insulin pump, an airbag — and you will find two people doing two different jobs, and you will find that the company is very clear about which is which.

The first person never touches the finished product. Their job is the *system*: the procedures, the calibrated machines, the training, the checklists, the standards that govern how the thing gets made. They work before the product exists and while it is being built. They practice **Quality Assurance**, and their goal is to make a defect nearly impossible in the first place.

The second person inspects what comes off the line. They stress the bridge, scan the device for cracks, pull the tenth unit and test it until it breaks. They work *after* the product exists, and their goal is to catch the defect that slipped through before it reaches a customer. They practice **Quality Control**.

Every serious maker of important things understands that you need both, and that they are not the same. Assurance is proactive and builds the system; control is reactive and inspects the output.

Confuse them — inspect harder instead of fixing the process, or trust the process and stop inspecting — and quality collapses in one of two predictable ways.

Here is the strange thing. We built this discipline for our products. We built it for our software, our food, our buildings, our drugs. We never built it for the most consequential thing any organization produces.

Manufacturing earned Quality Assurance. Software earned Quality Assurance. Your decisions — the most expensive things your enterprise makes — earned nothing.

Two jobs, one confusion

Think about how a consequential decision actually gets made in a large organization. A proposal arrives at a board — an architecture board, an investment committee, a change board, a credit committee. People examine it. They debate. Eventually someone, or some quorum, renders a verdict: approved, rejected, approved with conditions, escalated.

That final act — the verdict — is **Decision Control**. It is the gate. It is reactive: it responds to a specific proposal. It is decisive: it produces a yes or a no. And it is aimed entirely at the individual output. Decision Control is the human cousin of Quality Control — the inspection-and-verdict on one specific thing, the net that catches the bad one before it proceeds. It is also the older and better-credentialed of our two terms: the economists Eugene Fama and Michael Jensen wrote about decision control forty years ago — the ratify-and-monitor half of how organizations govern themselves. We did not invent it. We are pairing it with something new.

Because almost everything we built for decisions is Control. The org chart says who decides. The delegation-of-authority matrix says who may approve what. The governance committee exists to render verdicts. We invested centuries of organizational design in the question *who gets to make the call* — and almost nothing in a different question entirely:

Was the call made well — and can we prove it?

That second question is **Decision Assurance**. It is the system that makes sure the decision-making itself is sound: that the right criteria were applied, that nothing material was missed, that the rules that can never be broken were in fact not broken, that the reasoning rested on real evidence, that the dissent was heard and recorded, and that the whole thing left a trail you could hand to an auditor, a regulator, or your own future self. Decision Assurance is proactive. It is process-oriented. It works before and during the decision, not after. It is, precisely, the Quality Assurance of decisions.

We spent everything on Control and almost nothing on Assurance. And then we act surprised when our decisions turn out inconsistent, undocumented, dependent on who happened to be in the room, and impossible to defend six months later when someone asks why.

Why we keep reaching for more control

If Assurance is so obviously necessary, why did every organization build mountains of Control and almost none of it? The plain answer is that Control is *visible* and Assurance is *invisible*, and we are creatures who reward what we can see.

A new gate is a thing you can point to. Stand up a review board, add an approval step, require a second signature — and you can announce it, put it on a slide, show the auditor you *did something*. It feels like progress, and it is the move every anxious organization reaches for when a decision goes wrong: another checkpoint, another committee, another form. You met the hall of doors back in Chapter 1 — five gates and a three-month wait — and that scene is exactly what happens when an organization keeps answering an assurance problem with more control. Those decisions were not under-gated. They were under-assured. Piling on gates without assurance does not make decisions sounder; it just makes them slower, and leaves the proponent in the same fog, now standing in a longer line.

Assurance, by contrast, is quiet. A decision that was made completely, consistently, and on the record does not look like anything from the outside. It announces itself on exactly one occasion: the day someone asks "how do you know?" — and you, alone in the room, can give the answer.

What "made well" actually means

It helps to make "made well" concrete, because it is not a feeling. A decision is *assured* when five things are true of it — and you can show each one.

Here they are, one at a time. I wrote each so you could hold a real decision of your own up against it — because a list like this is worth keeping, and worth grading yourself against.

Complete. A decision is complete when every criterion that *should* bear on it actually did — when nothing material was skipped because the clock was running or because no one in the room owned that angle. The security review that never happened, the cost model no one built, the downstream team nobody consulted: each is a hole in completeness, invisible at the moment of decision and obvious in the post-mortem. To grade your own completeness, ask of a recent decision: is there a defined list of what *must* be considered for a decision of this kind — and can you show that each item on it was, in fact, considered? If the list itself does not exist, you are not scoring low on completeness; you are not measuring it at all.

Consistent. A decision is consistent when the standard that governs it is the same standard that governed the last one like it — when the answer does not swing with the calendar, the chair, or the mood of the room. The proposal that sails through this quarter on terms that sank an identical one last quarter is the signature failure here, and it corrodes trust faster than an outright bad call, because people feel the unfairness even when they cannot name it. To grade consistency, take two similar decisions from different months and different people, and ask whether the same rule visibly produced both. If the outcomes diverge and no one can point to *why* in the criteria, your decisions are being made by circumstance, not by standard.

Governed. A decision is governed when the rules that must never break did not break — and when that guarantee comes from the system, not from someone happening to remember it. There is a difference between a standard people *try* to honor and a constraint the process will not *let* them violate. The end-of-life technology that received an exception no rule actually permitted, the spending that quietly cleared its own ceiling: these are governance failures, the inviolable treated as merely advisory. To grade governance, name the handful of constraints in your domain that should be truly non-negotiable, then ask what would actually stop a determined, senior person from overriding one. If the only honest answer is "nothing but goodwill," the rule is a suggestion wearing a uniform.

Grounded. A decision is grounded when its reasoning rests on evidence a stranger could check — a number, a source, a document, a tested assumption — rather than on confidence, seniority, or a good story well told. This is the property the age of fluent AI makes most urgent, because a confident, fabricated rationale now costs nothing to produce. The vendor chosen on "a strong feeling about these people," the estimate that traces back to no model, the claim everyone repeated until it felt true: each is a decision floating free of anything verifiable. To grade grounding, take the central claim a decision rested on and try to follow it to its source. If the trail ends at someone's assertion rather than at something you could independently confirm, the decision was *asserted*, not grounded.

On the record. A decision is on the record when the whole of it — the verdict, yes, but also the options weighed, the reasons, the evidence, and the dissent — is captured somewhere durable that a stranger could find and follow months later. The reason that lives only in a meeting no one minuted, or in a direct message that vanishes when its author changes jobs, is a decision the organization did not actually keep. To grade this one, pick a consequential decision from a year ago and try to reconstruct *why* it was made, using only what was written down at the time. How far do you get before the trail goes cold? That distance is the exact measure of your institutional memory — and, usually, a sobering one.

Notice that not one of the five is about whether the decision turned out *right*. Assurance does not promise a correct outcome; nothing honest can. It promises a sound *process*, provably followed — which is the most any serious discipline ever promised, and far more than we offer our decisions today.

When the number is nobody's

Watch how grounding fails in the one place you would swear it could not — inside a number.

In 2012, a trading desk in the London office of one of the world's largest banks built a position so enormous that the trader behind it earned a nickname in the market: the London Whale. What let the position grow was not recklessness, exactly. It was a number. The bank measured the risk of its trades with a model that produced a single figure — how much the desk could lose on a bad day — and that figure stayed comfortingly low even as the position swelled. The risk managers watched the number. The number said fine. So the position grew.

The number was wrong, and the reason is almost too ordinary to believe. The model behind it was rebuilt in a spreadsheet, and the spreadsheet carried an error: at one step, a figure was divided by a sum where it should be divided by an average, and the mistake quietly cut the measured risk roughly in half. The risk the managers thought they were running was, in reality, twice as large. No one caught it, because no one independently checked the model that produced the most important number on the desk. It looked grounded — it was a number, from a model, on a screen. It was not. It rested on a spreadsheet nobody validated.

When the position finally turned, the loss reached about $6.2 billion. Afterward, the bank's own review and a United States Senate investigation traced the failure, in part, to that unverified spreadsheet — a single formula, trusted by everyone and audited by no one.

That is the trap the grounded property exists to catch, and it is subtler than a missing number. A number can be present, precise, and on every dashboard, and still be ungrounded — because grounding is not whether a figure exists. It is whether anyone followed the figure back to its source and confirmed the source holds. The London Whale's risk number was asserted with a model's authority and grounded in nothing anyone checked. The whole discipline of grounding is the refusal to trust a number simply because it is a number.

When the rule is only a wish

Two of those properties — *governed* and *on the record* — sound abstract until you watch them fail. So watch them fail, in two of the most expensive lapses of the last decade, both of which began with a rule that everyone agreed on and no system enforced.

Start on the morning of August 1, 2012, at Knight Capital — then one of the largest trading firms in America, an operation that handled around a tenth of all trading in U.S. stocks on a busy day.

Knight was deploying new code to its servers, a routine release, the sort the firm did all the time. There were eight servers to update. A technician updated seven. The eighth, for reasons no one caught, kept running old software — and the new release happened to repurpose a long-dormant flag that, on that one forgotten server, switched ancient, obsolete trading logic back on. Nothing in the deployment process confirmed that all eight machines matched. Nothing required a second set of eyes to sign off. And, fatally, nothing stood ready to halt a runaway.

At 9:30 that morning the opening bell rang, and the forgotten server woke up and began to trade — wildly, senselessly, buying high and selling low, firing millions of unintended orders into the market faster than any human could read them. The engineers in the room knew something was catastrophically wrong within minutes. What they lacked was a switch that simply said *stop*. They scrambled, and in the confusion one attempted fix made the bleeding worse. For roughly forty-five minutes the machine hemorrhaged money — about four hundred forty million dollars, gone, in the time it takes to eat lunch. Knight, a firm that woke that morning worth more than a billion, was effectively destroyed by lunchtime; it took emergency financing within days and was swallowed by a competitor within months. The Securities and Exchange Commission later cited it for violating the Market Access Rule — for, in plain terms, lacking the controls that would make such a disaster impossible.

Here is the assurance reading, and it is not a lesson about carelessness. The rules that mattered — *every server must match before release; a runaway must be stoppable* — were not missing. Everyone knew them. They were simply advisory: things a careful person tries to honor, rather than constraints the system will not permit anyone to violate. That is the exact gap the *governed* property closes. And notice the second lesson, quietly riding underneath: deploying code to a machine that moves real money at market speed is not a routine, two-way-door decision at all. It is a one-way door of the gravest kind, wearing the costume of a Tuesday release — a mis-leveling we will learn to name precisely in a later chapter.

Now hold that beside Equifax, where the rule was not just known but written down — and still went unenforced. In March 2017, a critical vulnerability in a widely used piece of web software called Apache Struts was disclosed to the world, and a patch was released the same day. Equifax — a company holding the financial intimacies of nearly every adult in America — did the responsible thing on paper: it circulated an internal order to apply the patch. The order went out. And then, on at least one public-facing consumer portal, no one applied it — and the automated scan meant to catch exactly that kind of miss failed to flag the exposed system. So the hole stayed open. Attackers found it, slipped through, and sat inside Equifax's systems, undetected, for roughly seventy-six days, quietly carrying off the names, Social Security numbers, birth dates, and addresses of about one hundred forty-seven million people — close to half the population of the United States. The Government Accountability Office later traced the catastrophe not to a single mistake but to a chain of missing controls, any one of which could catch it. Equifax eventually agreed to a settlement worth up to seven hundred million dollars.

The decision, in Equifax's case, was correct. *Patch the system* — the right call, made and documented. What failed was everything that should confirm the call was actually carried out. No control verified the order was executed. No record raised a hand to say one portal still stood exposed. The rule lived on paper and died in practice, and no one knew until a stranger was already inside. That is *governance* and *on the record* failing together — the whole difference between issuing a rule and proving, every time and for the record, that the rule was in fact followed. A decision you cannot show was carried out is, in the end, a decision you only hoped you made.

What we actually mean by a sound decision

We toss the phrase around constantly — *that was a sound decision* — and almost never stop to ask what it means. Try it on yourself for a moment. When you call a decision sound, do you mean it was well-informed? Well-reasoned? That it turned out well? Those are three different claims, and the slippage between them causes more confusion in governance than almost anything else.

Here is the trap, and a poker champion named it best. Annie Duke calls it *resulting*: the very human habit of judging the quality of a decision by the quality of its outcome. The investor who researched a company carefully and lost everything to an earthquake no one could foresee made a *good decision* with a *bad outcome*. The one who bought a stock on a whim and tripled their money made a *bad decision* with a *good outcome*. Outcome and decision quality are only loosely related, because the world runs partly on luck. Judge a decision by its result and you will praise recklessness whenever it pays and punish wisdom whenever it does not — and you will learn exactly the wrong lessons.

So let me be precise about the word, because our whole framework depends on it. In this book, a **sound decision is a well-*made* decision — an assured one.** It is sound if it was complete, consistent, governed, grounded, and on the record, *whatever it later produced*. Soundness lives entirely on the *making* side: it describes how the call was reached, not how the dice fell afterward. A sound decision can still end in failure, when luck turns against a good process — and that decision stays sound, and worth repeating. An unsound decision can still end in triumph, when luck rescues a careless one — and it stays unsound, and worth never repeating.

That is exactly why Decision Assurance assures the *making* and refuses to promise the *outcome.* It cannot make your luck good. It can make your decisions sound — and then make that soundness *provable,* so that when a good decision meets a bad result, you can show, to the auditor and to yourself, that the call was right even though the world was not. That is not a consolation prize. In a world you do not control, a provably sound process is the only thing you ever truly own.

Why you cannot skip either one

It is tempting, once you see the distinction, to think Assurance could replace Control — that a good enough system could simply make the decisions and dispense with the fallible human at the gate. Hold that thought; we tested it directly in the last chapter, and you saw how it went. For now, the principle.

Control without Assurance is judgment with no system behind it — a verdict you cannot examine, reproduce, or defend. It feels authoritative and is, in fact, fragile. The moment someone competent asks "how do you know?", it goes quiet.

I watched this happen to a capable executive once. A decision from the prior year — a platform his teams built a great deal upon — came under scrutiny when the platform faltered. The control worked perfectly at the time: the proposal went to the right board, the board voted, the minutes recorded an approval. But when the question came — *why this platform, what else was considered, who raised concerns* — the approval was all that remained. The criteria were never written down. The alternatives were a memory. The one architect who, people vaguely recalled, raised reservations was now at another company. The decision was controlled. It was not assured. And so a competent person, who very likely made a reasonable call with the information in front of him, stood before a room and could not prove it. That is the exact difference between a verdict and a *defensible* verdict, and it is worth a career to understand it before the morning you need it.

Assurance without Control is the opposite failure — a flawless process with no one accountable for the call. Process does not absolve; somebody must own the decision. You see this failure in the organization that builds a beautiful decision process — templates, criteria, a spotless record — and then lets the process itself become the decider, so that when something goes wrong, every individual points at the system. "I followed the procedure." "The form was complete." "The committee approved." A perfect trail leading to no one. Assurance can tell you a decision was made well; it cannot, and must not, tell you who to hold responsible for it. That is Control's job, and it is why the two never collapse into one.

The two are complementary, exactly as Quality Assurance and Quality Control are complementary, and the relationship between them is the key to everything that follows: **assurance is what makes control trustworthy.** A decision-assurance system does not take the decision away from the people responsible for it. It makes sure those people see everything relevant, apply the same rules every time, and leave a record they can stand behind — and then they make the call.

Stated as plainly as I can: a decision-assurance system **does not make the decision. It makes the decision defensible.**

That sentence will strike some readers as a limitation. It is the opposite. In an age when every vendor is promising a machine that will decide for you — and after a chapter that showed you what that machine actually delivers — the organizations that win will be the ones that got the division of labor right: they used the machine to *assure* their decisions and kept the *control*, the accountability, where it belongs.

And notice that the division of labor is about *roles*, not about humans versus machines. A human can assure — the careful reviewer who checks that every box was truly ticked is doing assurance work. A machine can control — the automated gate that auto-rejects an end-of-life technology is rendering a verdict. What never moves is the accountability. Whoever, or whatever, performs the assurance, and whoever, or whatever, renders the control, a human or an organization stays answerable for the decision. Assurance and Control are jobs to be done. Accountability is a person to be found.

So what?

For you, the distinction is a diagnostic you can run on your own organization tomorrow. Take everything you call "governance" and sort it into two piles. The verdict at the gate, the approval matrix, the committee that says yes or no — that is Control, and you almost certainly run plenty of it. Now look for the other pile: the system that guarantees each of those verdicts was complete, consistent, governed, grounded, and recorded. If that pile is thin or empty, you just found the gap this whole book is about — and you now own the word for it. You are not under-governed. You are over-controlled and under-assured.

What now?

Take one decision type that matters — a funding approval, an architecture sign-off, a credit call — and write down, plainly, which of the five assurance properties it can prove today. Complete? Consistent? Governed? Grounded? On the record? Most teams find they can claim one or two and only wish for the rest. That gap, named property by property, is your assurance backlog — and it is a far more useful list than "we should document things better," because each item tells you exactly what to build.

Where we go next

I keep calling assurance a *serious* word, and I owe you the proof. It is not a piece of jargon I coined to dress up a product. It is a discipline with a long and demanding pedigree, and in the most unforgiving corner of engineering — the software that keeps an airplane in the sky — it carries a precise, graded meaning that we are about to borrow wholesale. That lineage is the next chapter, and it is where Decision Assurance stops being an analogy and starts becoming a standard.

Chapter 4 in review

This section turns the chapter into study and teaching material — for self-assessment, for a graduate seminar, or as a train-the-trainer module.

Key takeaways

- Manufacturing split quality into two roles a century ago: **Quality Assurance** builds the system so defects rarely form; **Quality Control** inspects the output to catch the ones that slip through. You need both.
- Decisions deserve the same pair. **Decision Control** is the verdict at the gate — reactive, decision-by-decision (the older term, from Fama and Jensen, 1983). **Decision Assurance** is the proactive system that makes the *decision-making* sound. We over-built Control and under-built Assurance.
- A decision is **assured** when five things are true *and provable*: **complete, consistent, governed, grounded, on the record.**
- Assurance assures the *making*, not the outcome. A **sound** decision is a well-*made* one, independent of how it turned out. Judging a decision by its result instead of its process is the error Annie Duke calls **resulting**.

You cannot skip either job. Control without Assurance is a verdict you cannot defend; Assurance without Control is a process no one is accountable for. **Assurance is what makes Control trustworthy.**

The division of labor is about *roles*, not humans versus machines: a human can assure, a machine can control — but **accountability** never leaves a human or an organization.

The principle to carry forward: a decision-assurance system **does not make the decision; it makes the decision defensible.**

Key terms

Decision Assurance — the proactive discipline of ensuring, and being able to prove, that a decision was made completely, consistently, within its rules, and on the evidence; the Quality Assurance of decisions.

Decision Control — the reactive discipline of ratifying, rejecting, and monitoring an individual decision at the gate; the Quality Control of decisions (Fama and Jensen, 1983).

The five assurance properties — complete, consistent, governed, grounded, on the record.

Sound decision — a well-*made*, assured decision, whatever its outcome.

Resulting — the error of judging a decision's quality by its outcome's quality (Annie Duke).

Accountability (versus authorship) — answerability for a decision; it stays with a human or an organization even when a machine does the deciding.

Check your understanding

In one sentence, what is the difference between Decision Assurance and Decision Control?

Name the five properties of an assured decision.

True or false: a decision that led to a bad outcome cannot be a sound decision.

Who introduced the term *decision control*, and what does it pair with in their framework?

Finish the principle: "Assurance is what makes ______ trustworthy."

Answer key: (1) Assurance is the proactive system that makes the decision-making sound and provable; Control is the reactive verdict on a specific decision. (2) Complete, consistent, governed, grounded, on the record. (3) False — soundness describes how the decision was *made*, not how it turned out. (4) Fama and Jensen (1983); it pairs with *decision management* (initiation and implementation). (5) Control.

Questions for reflection and discussion

Of the five properties, which is weakest in your organization, and what would it take to make it *provable* rather than merely hoped for?

Recall a time a good decision got blamed for a bad outcome, or a reckless one got praised for a lucky win. How did "resulting" shape what people did next?

Name a constraint your organization treats as inviolable but enforces only by goodwill. What would change if the system, not a person, enforced it?

For facilitators: in sixty seconds, how would you explain assurance-versus-control to a skeptical executive who is sure they already run plenty of governance?

Apply it

Run the two-pile sort from this chapter. List your team's governance activities and sort each into **Control** or **Assurance**.

Then take one consequential decision type and grade it against the five properties, scoring each as *provable*, *partial*, or *absent*. The result is a one-page **assurance backlog** — a list of exactly what to build, in priority order. *Trainers:* run this as a thirty-minute small-group exercise, then debrief the property that came up emptiest across the room. It is almost always the same one.

Notes and further reading

Fama, E. F., and Jensen, M. C. (1983). "Separation of Ownership and Control." *(Decision management versus decision control.)*

Duke, A. (2018). *Thinking in Bets. (Resulting; decision quality versus outcome quality.)*

The Design Assurance Level idea this chapter foreshadows — aviation's DO-178C — is the subject of Chapter 5.

Canonical wording for the two terms: *DecisionAssurance_Canonical_Definition_v1.0.*

CHAPTER 5

The Lineage of Assurance

On a cold morning in October 1935, at Wright Field in Dayton, Ohio, the United States Army Air Corps gathered to watch the future of flight take off. The aircraft was Boeing's Model 299 — four engines where two was the norm, a wingspan over a hundred feet, more range and more bombs than anything in the sky. The contract was all but signed. Everyone expected a formality.

The plane taxied out, lifted off, climbed to three hundred feet, stalled, rolled onto one wing, and came down. The aircraft was lost. At the controls was no novice who needed another year of lessons. It was Major Ployer Hill, the Army's chief of flight testing — the most experienced flier in the corps. He simply forgot one step: a small lock that held the elevator controls still on the ground. With everything else this enormous machine demanded of him, one item slipped — and not even the finest pilot in the service could bring it back.

A newspaper called the Model 299 "too much airplane for one man to fly," and that was the moment the verdict could break two ways. The Air Corps could decide the plane was too complex and walk away. Or it could decide that the complexity was here to stay, and the human needed help carrying it.

They chose the second, and that choice is the reason you are reading this chapter. A group of test pilots sat down and did not ask for more training — their man owned all the training in the world. They wrote a list. Release the brakes. Close the windows. Unlock the controls. The most obvious actions imaginable, on a single index card, read aloud every time, by everyone, no matter how senior. With that list in the cockpit, the Model 299 went on to fly 1.8 million miles without a serious incident, and the Army bought thousands of them. You know the plane by the name it earned next: the B-17.

I open with this story because it is the whole book in miniature. A high-stakes output. A gifted expert. A silent, ordinary defect that no amount of skill reliably prevents. And a quiet system — not a smarter pilot, a *checklist* — that makes the outcome sound and provable, every time, for everyone. That is assurance. And aviation is only one of the places we already invented it.

Here is the part I want you to take to heart, especially if "Decision Assurance" still feels new, or invented. It is not new. It is one of the most thoroughly proven ideas your civilization owns. We worked it out, completely and at hard cost, at least four separate times — in the factory, in the network, in the ledger, and in the cockpit. Each time, the road was the same. Each time, the lesson was the same. I rushed past all four in a single breath back in Chapter 1; now let me slow down, because each one earned something we will need.

The factory: where assurance was born

We started here, so let me only add what Chapter 1 left out. When the factory finally tamed quality, it did so by splitting one word into two jobs. Quality Control is the inspector at the end of the line, pulling a finished part and testing it against spec — the verdict on the thing. Quality Assurance is the clipboard that watches the whole process, so the defect never forms in the first place — the system behind the thing. Control catches the bad part. Assurance builds the line that stops producing them.

W. Edwards Deming put the point as sharply as anyone ever put it. The third of his famous fourteen points for management is a single instruction: cease dependence on inspection. "Eliminate the need for inspection on a mass basis," he wrote in 1982, "by building quality into the product in the first place." You cannot inspect quality into a thing; you can only build it in. Inspection finds the bad part *after* you already paid to make it. Assurance changes the line so the bad part never gets made. Now hold that sentence against the way most organizations try to "improve" their decisions — another reviewer, another gate, another approval — and you will see the trap: we are still trying to inspect quality in. We pile control onto a process starved of assurance, and then act surprised when the defects keep coming.

That split — the verdict versus the system — is the one this whole book turns on, and the factory drew it first. Manufacturing earned Quality Assurance. Decision-making earned nothing of the kind. Everything that follows is us collecting a debt the factory paid off a century ago.

The network: assurance for the invisible

When information became the asset everything else rested on, we found that locking the door was not enough. Locking the door is control — a gate, a password, a yes-or-no at the threshold. But how do you *prove* that the data inside is what it claims to be, that it reached the right person, that the sender cannot later deny sending it? So the field built a discipline it named, fittingly, Information Assurance, and gave it provable properties. The United States Department of Defense settled on five: confidentiality (only the right eyes see it), integrity (no one tampered with it), availability (it is there when you need it), authentication (the parties are who they claim to be), and non-repudiation (no one can later deny what they did).

Notice what those five are. They are not a wish list. Each is a property you can test, demonstrate, and stand behind — assurance you can show a stranger. The network learned, the way the factory learned, that for anything that matters you do not merely guard it; you make its soundness provable.

And the resemblance to what we are building is closer than a feeling. Information Assurance grew by extending an older, narrower core — the CIA triad, for confidentiality, integrity, and availability — and then adding the two pillars decisions need most. Authentication asks whether the inputs truly came from who they claim. Non-repudiation makes certain no one can later deny what they did — which is to say, it puts the act on the record in a form the actor cannot wave away. Set those beside the properties of an assured decision and the lines connect almost one to one: integrity is *grounding*, the claim resting on something real; non-repudiation is being *on the record*; authentication and availability are the *governance* that the right inputs, from the right people, were actually in the room. The network already built, for data, four-fifths of what we are about to build for the decision.

The ledger: assurance as a profession

Money got there earliest, and gave us the most honest version of all. An entire profession — audit — exists for one reason: so that a number on a page can be believed by someone who was nowhere near the room when it was written. That is assurance as a service, the whole job.

The profession even wears its method in its name. "Audit" comes from the Latin *audire*, to hear — because medieval auditors did not read the accounts, they *heard* them, recited aloud by the steward while the auditor listened for the lie. The whole apparatus of modern assurance began as one person making another say it out loud, on the record, to a witness who did not work for them. Independence was built in from the first day, and we will see it again before this chapter ends.

And here is the detail I love, because it keeps us humble. The auditor's standard does not promise certainty. Under the

International Standard on Auditing (ISA) 200, an audit gives *reasonable* assurance — explicitly defined as high, but not absolute — because the standard admits that no real process drives the risk of error to zero. The evidence, it says plainly, is persuasive rather than conclusive. Sit with that. The most established assurance profession on earth, trusted with the world's money, built its entire credibility on refusing to overclaim. It does not say *certain*. It says *reasonable — and here is my evidence*. We are going to make exactly that promise about decisions, and not a word more. It is the same promise this book made in Chapter 3, when the experiment came back humbling: assurance, not omniscience.

The cockpit: assurance, graded

Which brings us back to where we started, one level deeper. After the checklist saved the B-17, aviation kept going — all the way to the most demanding software discipline humans run. The software that flies a modern aircraft is certified against a standard with an unglamorous name, DO-178C, *Software Considerations in Airborne Systems and Equipment Certification*, and it does something the other three fields only imply. It grades the rigor to the stakes.

Here is how. Every piece of software on the aircraft is assigned a level — a Design Assurance Level — by asking one question: if this fails, how bad is it? If a failure would be catastrophic, the kind that loses the aircraft, the software is Level A, and it must satisfy the full battery of assurance objectives — by the common count, seventy-one of them, many checked by someone independent of the people who built it. If a failure would be hazardous, it is Level B, with sixty-nine. Major, Level C, sixty-two. Minor, Level D, twenty-six. And if the software failing changes nothing that matters — a cabin reading light flickers — it is Level E, and the standard asks for *nothing at all.*

Read that ladder again, because it is the most important idea in this chapter. Aviation does not assure everything to the same punishing degree. That would be impossible, and wasteful, and would bury the critical work under the trivial. Instead it spends its rigor where the consequences live. The flight-control computer whose failure would lose the aircraft gets seventy-one objectives. The reading light gets none. The rigor matches the consequence — precisely, deliberately, on a scale.

Two details lift this above a simple sliding scale, and both lead straight back to assurance. First, you do not choose your level; the consequence chooses it for you. The level falls out of a formal safety assessment that asks what a failure would actually do, so no team gets to talk its own risky component down to an easier grade. Second, the higher the level, the more of its objectives must be met *with independence* — checked by someone other than the person who did the work. That is the audit principle again, the witness who does not work for you, now written into software. The same idea we met in the ledger, and the same answer to the question Chapter 2 left ringing — who checks the checker? Someone independent, by design.

The one law under all four

Stand back and look at the four fields together, and the pattern stops looking like a coincidence.

Field	The silent defect that forced it	What they built	The recurring instrument
Manufacturing	recalls, defects, harm on the line	Quality Assurance — build quality in, do not inspect it in	the control chart, the process audit
Information	data too valuable to lose or forge	Information Assurance — five provable properties	the security control, the access log

Audit (finance)	numbers no outsider could trust	reasonable, not absolute, assurance — given independently	the audit trail, the independent opinion
Aviation	machines that outgrew the pilot	Design Assurance Levels — rigor graded to consequence	the checklist, the graded objective set

The same law is written across every row.

> *When the cost of a silent defect grows too high to bear, every serious field stops trusting the talented individual and builds a system that makes soundness provable — and it scales the rigor to the stakes.*

That is the lineage of assurance, in a sentence. It is not four good ideas that happen to rhyme. It is one idea, discovered four times, because it is true. The factory found it when the recalls and the harm piled up. The network found it when data became too valuable to lose. The ledger found it when money became too consequential to take on trust. The cockpit found it when the machines outgrew the pilots. Four roads, one destination — and the same two tools waiting at the end of each: an instrument you can hold (a checklist, a log, a trail) and an outsider who checks the work.

The one place we never went

So here is the question that should be making you a little uncomfortable, because it is the one that started this book. We assure the bolt. We assure the byte. We assure the books. We assure the flight. We do not assure the *decision* — the single highest-consequence thing an enterprise produces, the thing that sets all the bolts and bytes and flights in motion. It is the one expensive output we still hand to a talented individual and hope.

Decision Assurance is simply us walking the road the other four already walked, for the output we somehow skipped. And the cockpit hands us the most useful gift of all: the graded ladder. We do not need to assure every decision to seventy-one objectives. A reversible, five-hundred-dollar call does not deserve the scrutiny of a billion-dollar, one-way door — and burying the second under the paperwork of the first is how assurance dies of its own weight. The answer aviation already worked out is to grade it. Match the rigor to the consequence. That idea — a Decision Assurance Level for the decision, set by what happens if the decision is silently wrong — is the spine of the chapters ahead, where it grows from a metaphor into a working scale of A through E.

A first checklist: match the rigor to the consequence

Since this is the chapter where the graded ladder enters, here is the first tool to take to your desk. It does for a decision what the Design Assurance Level does for a line of flight software: it sets the level of assurance by the size of the consequence. (And the guard that rides on every checklist in this book: it prompts a decision, it never replaces one. It assures that you thought; it does not think for you.)

Ask, of a decision in front of you:

If this turns out silently wrong, how bad is it? Catastrophic (you cannot recover), severe, major, minor, or no real effect?

Is it reversible? A door you can walk back through needs less assurance than a one-way door.

How much does it bind — how many people, how much money, for how long?

So what level of assurance does it deserve — the full weight of the five properties, or a light touch — and are you, right now, spending more on the trivial than on the grave?

Most of us, run carefully through those four lines, discover the same thing: we over-assure the small reversible decisions that generate paperwork, and under-assure the enormous one-way doors that generate consequences. Naming the level is how you stop. *(This is the Field Kit's Checklist 4; it lives there too, for the day you want it without the chapter around it.)*

So what?

Assurance is not a novelty I am asking you to trust. It is a proven discipline your own world already runs — four times over — and you already rely on it every time you board a plane or believe an audited number. The only news here is *where* we point it. And the cockpit's graded ladder means Decision Assurance asks for judgment, not zealotry: you spend rigor where the stakes are, and nowhere else.

What now?

Take one real decision from the past month and give it a level. Ask the first checklist question — if this were silently wrong, how bad would it be? — and then ask whether the care you actually put in matched the answer. If you assured a reversible memo to the hilt and waved through an irreversible commitment on a good feeling, you just found the gap, and the rest of this book is how you close it.

Where we go next

We now hold the law — rigor matches consequence — and the shape of the instrument: a graded level, set by stakes, checked by someone independent. What we still lack is the machinery — what the levels actually are for a decision, what each one demands, and how a real organization runs them without grinding to a halt. That is the next chapter, where the lineage becomes a framework.

Chapter 5 in review

Key takeaways

Assurance is **not new and not mine** — civilization built it, fully and at hard cost, at least four times: in manufacturing, information, audit, and aviation.

Manufacturing split quality into Control (inspect the finished part) and Assurance (build the process so the defect never forms). As Deming insisted: you cannot inspect quality in — you build it in. The same is true of decisions.

Information Assurance gave the invisible asset five provable properties — confidentiality, integrity, availability, authentication, non-repudiation — that map almost one-to-one onto the properties of an assured decision.

Audit is assurance as a profession, and the most honest of all: it promises *reasonable* assurance, "high but not absolute," resting on evidence that is persuasive, not conclusive — and given *independently*.

Aviation added the masterstroke — it **grades rigor to stakes**. Under DO-178C, the most catastrophic software meets the full battery of objectives; software that cannot affect safety meets none. The consequence picks the level, not the team.

The one law under all four: **when a silent defect becomes too costly to bear, a serious field stops trusting the individual, builds a system that makes soundness provable, and scales the rigor to the consequence** — with an independent check at the high end.

The decision is the **one high-consequence output we never put through this** — and the graded ladder becomes the **Decision Assurance Level**, the spine of the chapters ahead.

Key terms

Quality Assurance / Quality Control — building the process so defects cannot form (assurance) versus inspecting the finished item (control); "build quality in, do not inspect it in" (Deming).

Information Assurance — the discipline of making information's soundness provable; the United States Department of Defense names five pillars: confidentiality, integrity, availability, authentication, non-repudiation.

Reasonable assurance — in audit, a high but deliberately *not absolute* level of assurance; the honest admission that no real process drives risk to zero.

Independence — the assurance principle that the checker must not be the maker: the auditor who does not work for you, the DO-178C objectives met "with independence." The standing answer to "who checks the checker?"

DO-178C — *Software Considerations in Airborne Systems and Equipment Certification*, the standard governing airborne software.

Design Assurance Level (DAL) — the graded level of rigor assigned to a component by the severity of its failure; in DO-178C, software levels A (catastrophic) through E (no effect).

Decision Assurance Level — our adaptation: the level of assurance a decision deserves, set by what happens if the decision is silently wrong.

Check your understanding

Name the four fields that independently invented assurance, and the one law they share.

What is the difference between Quality Assurance and Quality Control, and what did Deming say about inspection?

Why does the audit profession promise *reasonable* rather than *absolute* assurance — and why do we borrow that humility?

In DO-178C, why does Level A carry far more objectives than Level D, and who decides a component's level?

What single mechanism recurs in both the audit profession and DO-178C's higher levels?

What is a Decision Assurance Level, and what sets it?

Answer key: (1) Manufacturing, information, audit, aviation; the law is that when a silent defect grows too costly, a field stops trusting the individual, builds a system that makes soundness provable, and scales rigor to the stakes. (2) Assurance builds the process so defects never form; Control inspects the finished item — and Deming's third point is to cease dependence on inspection and build quality in. (3) Because no real process drives the risk of error to zero, so honesty demands "high but not absolute"; we borrow it because Decision Assurance promises a sound, provable *process*, not a guaranteed outcome. (4) Because a Level A failure is catastrophic and a Level D failure is minor; the level is set by a formal safety assessment of the consequence, not chosen by the team. (5) Independence — the checker is not the maker. (6) The level of assurance a decision deserves, set by the severity and reversibility of being silently wrong.

Questions for reflection and discussion

Which of the four lineages is closest to your own profession, and what did it teach you to assure that you do not yet assure in your decisions?

Where in your organization do you spend the most assurance on the least consequential decisions — and the least on the most?

Where today does the checker also happen to be the maker — and what would real independence cost you?

For facilitators: ask the room to name a decision that deserves "Level A" treatment and one that deserves "Level E." The argument over which is which is where the learning lives.

Apply it

Take five real decisions your team made last quarter, across a range of sizes. For each, run the chapter's first checklist: how bad if silently wrong, how reversible, how much it binds. Sort the five into assurance levels — call them A through E for now. Then mark, without flattery, how much rigor each one actually received. The mismatches — the Level A call made on instinct, the Level D memo over-engineered to death — are your first map of where assurance is mis-allocated. *Trainers:* run the sort in small groups and compare where teams disagree on a decision's level; the disagreement itself teaches the framework.

Notes and further reading

The Boeing Model 299 / B-17 checklist origin is recounted in Atul Gawande, *The Checklist Manifesto* (2009); the 30 October 1935 Wright Field crash, the forgotten control lock, and the 1.8-million-mile record are widely documented.

W. Edwards Deming, *Out of the Crisis* (1982); the third of the Fourteen Points: "Cease dependence on inspection to achieve quality… build quality into the product in the first place."

DO-178C, *Software Considerations in Airborne Systems and Equipment Certification* (RTCA; EUROCAE ED-12C). Software levels A–E and the commonly cited objective counts (71 / 69 / 62 / 26 / 0) appear across DO-178C references. The term *Design Assurance Level* is formalized at the system level in SAE ARP4754A (which calls it an Item Development Assurance Level); DO-178C itself uses "software level."

The five pillars of Information Assurance are the United States Department of Defense model; they extend the older CIA triad (confidentiality, integrity, availability).

Reasonable assurance is defined in International Standard on Auditing (ISA) 200. "Audit" derives from the Latin *audire*, "to hear."

The Quality Assurance / Quality Control distinction and its manufacturing lineage (Shewhart, Deming) are introduced in Chapter 1; the five assurance properties are defined in Chapter 4.

PART III

The Framework

CHAPTER 6

How Much Assurance Is Enough?

Somewhere in your organization, right now, a capable adult is filling out a three-page form to requisition a second monitor. Two floors up, someone just approved a two-million-dollar platform on the strength of a confident slide and a nod in a hallway. Hold those two pictures side by side, because together they are the most common assurance failure in the modern enterprise — and notice that the failure is not *too little* assurance. It is assurance aimed at the wrong target. We lavish scrutiny on the reversible and the trivial, and we wave the irreversible and the enormous straight through.

Jeff Bezos named half of this in 2015. Some decisions, he wrote, are one-way doors — consequential and nearly impossible to reverse; walk through, dislike what you find, and you cannot get back. Those deserve to be made slowly, carefully, with consultation. Most decisions, though, are two-way doors: reversible, cheap to undo, and best made fast by good people who do not wait for certainty. And then the warning that should sound familiar after Chapter 1: as organizations grow, they tend to apply the heavyweight, one-way-door process to *everything* — including the two-way doors. The result is the monitor form and the rubber-stamped platform, side by side.

Bezos was right, and he stopped one step short of the tool. Knowing a decision is a one-way door tells you to slow down. It does not tell you *what assurance to bring* — which of the five properties to prove, how much independence to demand, whether to fly the simulator, whether to call an audit. That is what this chapter gives you: a graded scale, borrowed straight from the cockpit, that turns "how big is this?" into "here is exactly how much assurance it earns."

The two axes

Recall the cockpit from Chapter 5. Aviation does not assure every line of flight software to the same punishing degree; it assigns each component a Design Assurance Level by the severity of its failure, and spends seventy-one objectives on the catastrophic and none on the harmless. We do the same for decisions — and we set the level on two axes, not one:

Severity — how bad is it if this decision is silently wrong? From existential at the top down to no real effect at the bottom.

Reversibility — one-way door or two-way door? Can you walk it back, and at what cost?

The level is set by the *worse* of the two. A decision that looks moderate in severity but slams an irreversible door behind it is not moderate; the door makes it grave. And here is the rule that keeps everyone honest: the consequence assigns the level, not the person who would prefer an easy ride. You do not get to talk your one-way door down to a two-way door because the scrutiny is inconvenient. The assignment, like the aviation safety assessment, sits above the decider — which is the independence principle from Chapter 5, applied before the decision rather than after.

The five Decision Assurance Levels

Read the right-hand column as a single dial. One setting governs how much of *everything* — properties, independence, simulation, audit — a decision earns.

Level	The decision	What assurance it earns
A — Existential	A one-way door that could end the enterprise or cause irreversible harm: betting the company, an irreversible safety, compliance, or ethical call.	All five properties, fully proven. An independent Decision Audit and a Seal of Assurance. The Decision Flight Simulator flown forward before you commit. Decision and expected outcome pre-registered. Dissent recorded.
B — Severe	A heavy door, costly and hard to reverse: a major strategic commitment, a top hire, a market entry, large sunk spend.	All five properties proven. Independent review — a second set of eyes the decider does not own. A simulator forward pass. Rationale, evidence, and dissent on the record.
C — Major	A door that costs real money or time to reopen: a vendor selection, an architecture standard, a sizable budget allocation.	The five properties present and recorded. A grounding-and-completeness checklist. Logged in the ledger. No formal audit — but a process you could seal.
D — Routine	A two-way door, quick and cheap to undo: most day-to-day approvals, tooling choices, small allocations.	A light touch: record the decision and the why. Trust the assured process, not an audit of the instance. Decide fast.
E — Trivial	No real door at all: cosmetic, fully reversible,	Nothing. Decide and move on. Rigor spent here is

	no material effect.	pure waste.

A few words on each, because the whole framework lives in the calibration.

Level A is for the decisions you cannot take back and cannot survive getting wrong. Betting the company on a platform. An irreversible safety, environmental, or ethical call. A merger that cannot be unwound. Most organizations see only a handful of these a year, and that rarity is the point: Level A earns the full battery — every property proven, an independent auditor, a Seal, the Flight Simulator flown forward to surface the critical dependency before the door closes, and the expected outcome written down in advance so the decision can be learned from later. This is aviation's seventy-one objectives, translated into a boardroom.

Level B is the serious, expensive, hard-to-reverse call that stops short of existential. A major market entry, a senior hire, a multi-year commitment. It earns all five properties and a truly independent reviewer — not the apparatus of a full audit, but a set of eyes the decider does not control — plus a forward run of the simulator and a complete record, dissent included.

Level C is the workhorse of real governance — the vendor, the standard, the budget. It earns the five properties, captured and checklisted, and a home in the ledger. No formal audit rides on it, but the process behind it should be sound enough that an auditor *would* seal it if asked. Most of what an Architecture Review Board touches lives here.

Level D is where most decisions actually live — the two-way doors. Here the discipline inverts: the failure is not under-assuring, it is over-assuring. A Level D decision earns a light record — what was decided and why — and otherwise trusts the assured process around it rather than auditing the single instance.

Decide it quickly, with good people, and move. This is exactly the door Bezos warned us not to barricade.

Level E asks for nothing, and meaning it is part of the discipline. The cosmetic, the fully reversible, the immaterial. Spending assurance here is not virtue; it is the monitor form. The courage to assure *nothing* — to let a trivial decision simply be made — is as much a part of mature assurance as the rigor at the top.

When the stakes and the assurance drift apart

The Levels exist to keep two things lashed together — the consequence of a decision and the rigor brought to it. The quiet catastrophe is what happens when they drift apart: when something operating at Level-A stakes is assured as if it were Level E. To see how easily that happens, follow a machine into a place where the stakes could hardly be higher — the examining room.

Transcribing a doctor's visit is tedious, time-consuming work, and clinicians drown in it, so the arrival of AI transcription felt like deliverance. Tools built on OpenAI's Whisper model spread quickly through hospitals and clinics, quietly turning the spoken visit into the written record. One vendor's Whisper-based tool alone reached tens of thousands of clinicians across an estimated seven million medical visits. A godsend — until you learn what Whisper sometimes does.

Like every model of its kind, Whisper occasionally invents. It does not merely mishear; it fabricates — adding whole sentences that no one in the room ever said, including invented medical content, slipped seamlessly into the transcript in the same matter-of-fact tone as the real words. An Associated Press investigation found the behavior widespread, with researchers placing the hallucination rate anywhere from around one percent of transcripts to a majority, depending on the study. And here is the detail that turns a flaw into a genuine danger: the tool deletes the original audio once the transcript is made. The recording — the one piece of ground truth that could prove what the patient actually said — is gone. So a sentence the machine invented becomes a permanent line in a medical record, indistinguishable from the words a human spoke, with nothing left to check it against.

Sit with the level mismatch. A medical record is about as close to a Level-A artifact as ordinary life produces: it drives diagnoses, prescriptions, surgeries — decisions you cannot take back and cannot survive getting badly wrong. The stakes demand the full battery: grounding every line to its source, a verifiable record, an independent check. What the technology delivered instead was the opposite of all three — an ungrounded machine inventing content, then destroying the only evidence that could ground it. Level-A stakes, met with Level-E assurance. The point of grading is to make sure that mismatch never happens by accident: that the gravity of the decision, not the convenience of the tool, sets how hard the assurance works. A record you cannot check against its source is not assurance. It is a rumor with a timestamp.

A decision, leveled

Frameworks persuade in the abstract and fail in the particular, so let me walk one decision all the way through.

Imagine your team wants to standardize on a new core platform — a vendor that will sit underneath a dozen systems. On the first pass it looks like a Level C: a vendor selection, the workhorse of governance, reversible at some cost. Someone drafts the criteria, the properties get recorded, and it heads toward a routine approval.

Then you run the two axes properly. Severity is real but not existential — call it major. Reversibility is where the surprise lives. "We can always switch later," the team says, and on the contract's face that is true. But the forward pass of the Flight Simulator — the dependency graph projected a year out — tells a different story. Three downstream decisions are already queued to build directly on this platform within two quarters, and the contract carries a multi-year data-egress penalty. Put those together and the honest reading is plain: eighteen months from now, this is no longer a two-way door. The reversibility you are counting on expires quietly, long before anyone thinks to look for it.

That single finding re-grades the decision. The worse of the two axes wins, and the hidden one-way door pulls it up from C to B. Now the B treatment applies — a truly independent reviewer, the simulator run on the record, the rationale and the dissent captured — and the team makes the same choice, or a different one, with its eyes open. Either way, the decision is now defensible.

Notice what just happened, because it is the whole framework working at once. The level was not a label slapped on at the start; it was *discovered*, by looking at the decision's dependencies before committing to it. Grade it C and stamp it, and you would meet that egress penalty the way organizations usually do — eighteen months later, in a budget meeting, with no memory of who agreed to it or why. The level is how you find the one-way door while you can still choose not to walk through it.

Five decisions, five levels

One decision in depth shows the mechanism; five at a glance show the range. Here are five real calls an enterprise makes, run through the same two axes, landing across the whole scale.

The decision	Severity	Reversibility	Level (worse of the two)
Make one vendor the sole processor of all customer payments, on a five-year exclusive with a punishing	Severe-to-existential	One-way door	A

exit			
Enter a new country, standing up a local entity and eighteen months of sunk investment	Severe	Heavy door	B
Choose the standard analytics platform for the data organization	Major	Reversible at real cost	C
Adopt a new code-formatting tool for one team	Minor	Two-way door	D
Change the color of an internal dashboard header	None	Fully reversible	E

Read down the table and the scale does its work: the full battery at the top, nothing at all at the bottom, and a graded slope between. But the row that teaches the most is the first one — because on severity alone, a *vendor contract* sounds like a Level C, the workhorse tier where most procurement lives. What lifts it four levels is not its severity. It is the door. A five-year exclusive that processes every dollar of revenue, with an exit penalty designed to trap you, is a one-way door welded to the company's survival. The worse-of-two rule sees that instantly; severity alone never would. Grade it on "it's just a vendor selection," and you would meet its true level eighteen months later, the first time the vendor raised its price and you discovered you could not leave.

That is the rule earning its keep in a single line: the worse of the two axes wins, and the axis that catches the dangerous decisions is almost always reversibility — because severity is loud and reversibility is quiet. A grave consequence announces itself. A closing door does not. The whole art of leveling is hearing the door.

One dial, not five separate spends

Notice what the level actually does. It is a single dial that sets every other safeguard at once. How deeply you prove the five properties, whether you bring an independent set of eyes, whether the Flight Simulator flies, whether a Decision Audit and a Seal apply, whether you pre-register the expected outcome — all of it scales with the level. That is the elegance the cockpit taught us. You do not negotiate each safeguard separately, decision by decision, until governance collapses under its own weight. You set the level, and the level sets the rest.

Two cautions, in the book's own spirit. First, a level is the assurance *effort*, not a guarantee. Level A means assured to the highest provable standard — reasonable, not absolute; even seventy-one objectives never promised a perfect flight, only a defensible one. Second, when you cannot tell whether a decision is a B or a C, default *up* one level — and then resist the gravity Bezos named, the pull to treat everything as Level A. The framework earns its keep in both directions: it licenses you to grade *down* the reversible many, precisely so you can afford to grade *up* the irreversible few.

Leveling at scale: rules, not rulings

One worry should be nagging at you by now. If every decision needs two axes, a checklist, and sometimes a simulator run, does the framework not become its own three-page monitor form — assurance overhead swallowing the very speed it promised to protect?

It would, if you leveled every decision by hand. So you do not. The trick is to encode the common cases as *rules*, not *rulings*. A ruling is a human weighing a single decision; a rule is a standing line that levels a whole class of them automatically. "Any spend above this figure, or any commitment we cannot exit within ninety days, is at least Level B." "Any change to a customer-facing safety control is Level A." "Any reversible, sub-threshold tooling choice is Level D — decide it and move." Written once, those lines level thousands of decisions without a meeting, and they reserve human judgment for the two places it is actually needed: the truly ambiguous, and the truly grave.

This is leveling as *governance-as-code* — the rules themselves are a governed artifact, versioned, owned, and assured like any other. And it inverts the overhead worry entirely. A mature leveling policy makes the organization *faster*, because the two-way doors stop waiting in the one-way-door queue. The framework does not add a gate; it removes the gates you were applying by reflex, and points the few that remain at the decisions that earn them.

A checklist: set the level

Before you spend a minute of assurance on a decision, spend thirty seconds setting its level. (The guard, as ever: this names the rigor a decision earns; it does not make the decision for you.)

How bad if silently wrong? Existential, severe, major, minor, or none.

One-way or two-way door? Can you reverse it, and at what cost?

Take the worse of the two. An irreversible door is never below Level B, whatever the severity feels like in the moment.

When you cannot tell, default up one level — then ask whether the reversible truth lets you settle back down.

Set the dial. The level you land on sets everything else: how much of the five properties to prove, how much independence to demand, whether the Flight Simulator flies, and whether a Decision Audit and Seal apply.

(This is the Field Kit's Checklist 4, now grown its second axis — reversibility — and the default-up rule.)

So what?

The blind spot in Chapter 1 was never that we assure too little in total. It is that we assure blindly — pouring scrutiny where it is cheap and skipping it where it is dear. The Decision Assurance Levels convert an impossible mandate, *assure everything*, into a workable one, *assure proportionally*. For the first time you can answer "how much assurance is enough?" with something other than a shrug or a rubber stamp: enough for the level, and not a form more.

What now?

Take your last ten real decisions and assign each a level with the checklist above. Then mark how much assurance each one actually received. The mismatches will sting in two directions: the one-way door you made like a two-way door, and the trivial call you tortured like an existential one. That two-column list — under-assured at the top, over-assured at the bottom — is your re-allocation plan, and it usually frees more capacity than it consumes.

Where we go next

A level raises an obvious question: *who* assigns it, who proves it, and who gets to place the Seal? Grading a decision is itself a kind of judgment, and judgment, this whole book insists, must be assured by someone independent of the decider. That points straight at a role we keep circling — the person with the clipboard, the one who audits the decision rather than making it. The next chapter gives that role a name, a body of knowledge, and a seal.

Chapter 6 in review

Key takeaways

The common failure is not too little assurance but **misallocated** assurance — heavy scrutiny on the reversible and trivial, a rubber stamp on the irreversible and grave.

Bezos named the reversibility half: one-way doors (consequential, irreversible) deserve deliberation; two-way doors (reversible) deserve speed — yet large organizations wrongly apply the heavyweight process to everything.

Set a decision's level on **two axes** — severity (how bad if silently wrong) and reversibility (one-way or two-way door) — and take the **worse** of the two.

The **consequence assigns the level, not the decider** — independence applied *before* the decision, the way aviation's safety assessment sets a component's level.

The five **Decision Assurance Levels** — A Existential, B Severe, C Major, D Routine, E Trivial — run from the full battery (audit, Seal, Flight Simulator, pre-registration) down to *nothing at all.*

The level is **one dial**: it sets the depth of the five properties, the independence, the simulation, and the audit all at once.

The framework licenses grading **down** the reversible many so you can afford to grade **up** the irreversible few — and a level is assurance *effort*, reasonable not absolute, never a guarantee.

The level is **discovered, not labeled**: a decision's dependencies can reveal hidden irreversibility that re-grades it upward (often surfaced by the Flight Simulator), and leveling at scale runs on **rules, not rulings** (governance-as-code), so the framework speeds the organization rather than slowing it.

Key terms

Decision Assurance Level (DAL) — the graded rigor a decision earns, set by its consequence; levels A (existential) through E (trivial), adapted from aviation's Design Assurance Level.

Severity — how damaging it is if a decision is silently wrong, from existential to no effect.

Reversibility (one-way / two-way door) — whether a decision can be walked back, and at what cost (Bezos, 2015); a one-way door raises the level regardless of apparent severity.

The dial — the principle that the level sets every other safeguard at once, rather than each being negotiated separately.

Default-up rule — when the level is unclear, choose the higher one, then grade back down only if reversibility allows.

Re-leveling — revising a decision's level when its dependencies reveal hidden irreversibility, frequently caught by the Flight Simulator's forward pass.

Leveling policy (governance-as-code) — assigning levels by standing rules rather than per-decision rulings, so most decisions self-level and human judgment is reserved for the ambiguous and the grave.

Check your understanding

What are the two axes that set a decision's level, and which one wins when they disagree?

Why does the *consequence* assign the level rather than the decider?

Name the five Decision Assurance Levels and the rough decision each fits.

What is the failure mode at Level D and E — and why is "assure nothing" sometimes correct?

What does it mean to call the level a single "dial"?

What can the Flight Simulator's forward pass reveal that changes a decision's level — and what is that called?

Answer key: (1) Severity and reversibility; the worse of the two sets the level (an irreversible door is never low). (2) To stop people talking a grave decision down to an easy one; independence applied before the decision. (3) A Existential (bet-the-company / irreversible), B Severe (major hard-to-reverse), C Major (vendor / standard / budget), D Routine (reversible day-to-day), E Trivial (cosmetic). (4) Over-assuring — drowning reversible, trivial decisions in process; assuring nothing at Level E frees capacity for the decisions that matter. (5) Setting the level sets the depth of the five properties, the independence, the simulation, and the audit all together. (6) Hidden irreversibility — downstream dependencies or lock-ins that turn an apparent two-way door into a one-way door, pulling the level up; that is re-leveling.

Questions for reflection and discussion

Pick a decision your team agonized over and one it rushed. With the two axes, were they graded correctly — or backward?

Where does your organization most apply the "one-way door" process to two-way-door decisions, and what does that cost in speed?

Where does a "reversible" decision in your world quietly become irreversible — and who, if anyone, would notice before the door shut?

For facilitators: ask the group to level the same five real decisions independently, then compare. The disagreements expose where the organization's risk appetite actually lives.

Apply it

Build your decision's level into the workflow, not your memory. Take one decision type your team handles often, and write a one-line rule that assigns its level automatically — for example, "any spend over X, or any vendor we cannot exit within ninety days, is Level B." Then attach the matching assurance to each level: what gets recorded, who reviews, when the simulator runs. You are turning the dial into policy, so no one re-litigates it per decision. *Trainers:* ask each participant to draft one such rule for their own domain and pressure-test it against a real decision that would trip it.

Notes and further reading

The one-way / two-way door (Type 1 / Type 2) framing, and the warning that large organizations over-apply the heavyweight process, are from Jeff Bezos's 2015 Amazon shareholder letter.

The graded-rigor principle and aviation's Design Assurance Levels (DO-178C) are established in Chapter 5; the five assurance properties in Chapter 4.

The Decision Flight Simulator, the two ledgers (pre-registering the expected outcome), and the Seal of Assurance are developed in the chapters that follow; this chapter only sets the dial that governs how much of each a decision earns. *Leveling policy as governance-as-code* is carried further in the implementation chapter.

CHAPTER 7

The Decision Auditor

Somebody must audit the auditor.

Why this job appears now

Every so often a new profession appears that did not exist a generation earlier — not a fresh title for old work, but an entirely new kind of expert the world suddenly cannot do without. The financial auditor was one. The information-systems auditor was another. The decision auditor is the next, and the question worth sitting with before we describe the role is the one a skeptic asks first: why now? If decisions matter this much, why did no one build a profession to assure them in 1995, or 1975?

The answer is a pattern, and once you see it you cannot unsee it. A new assurance profession is born at the moment a technology becomes two things at once: consequential enough that being wrong is intolerable, and auditable enough that being right can be proven. Not before. The profession always lags the technology by about a generation, because the world must first come to depend on the thing, and the tools to inspect it must first mature.

When a technology turned consequential and auditable…	…a profession was born	roughly
the modern corporation and public securities	the financial auditor / Certified Public Accountant	1930s
the business computer	the information-systems	1969

	auditor	
the internet and mass data	information security and privacy	2000s
artificial intelligence and autonomous decisions	the decision auditor	now

Look at the gaps. The computer arrived in the 1940s; its auditors organized in 1969, once computing carried enough of the business to need governing. We stand about that same distance from the arrival of capable artificial intelligence — and the two conditions just turned true for the decision itself.

On the supply side, the cost of the machine that can do this work fell off a cliff. The price of running a capable model dropped roughly two hundred and eighty-fold in eighteen months, and by one industry measure it is falling tenfold a year — faster than computing fell in the personal-computer revolution or bandwidth fell in the dotcom boom. And the deeper shift is not the price; it is the *kind* of machine. Every earlier attempt to assure a decision ran into the same wall: the only tools were brittle, hand-coded rules, and a decision does not live in rules. It lives in prose, in scattered evidence, in the dissent someone voiced and no one wrote down. For the first time, a machine can read that. The falling cost made it affordable; the new kind of model made it possible; both arrived together.

On the demand side, the same technology created the very problem it can now solve. As organizations hand decisions to artificial intelligence — the copilot, the model that grades, the agent that acts — consequential decisions multiply faster than any human governance can follow, and they arrive opaque. Artificial intelligence is, in the same moment, the flood and the ark: it produces decisions faster than we can answer for them, and it is the only thing fast enough to assure them. You did not need a decision auditor in 1995, because the machines were not yet deciding. Now they decide by the million, and no one can answer for them.

And the law is arriving on the same clock — Europe's artificial-intelligence rules, the data-protection right to a human in the loop, model-risk rules in banking — each demanding the auditable trail we never kept. Regulation is what turns a good idea into a profession, the way the securities laws of the 1930s turned bookkeeping into audit.

That clock is not hypothetical, and you do not need to wait for a statute to hear it ticking — the consequences are already landing, by the hundreds, in courtrooms. The fabricated-citation problem that began with one embarrassed lawyer became, within two years, a wave. By 2025 a California appeals court fined an attorney named Amir Mostafavi ten thousand dollars after finding that twenty-one of the twenty-three quotations in his brief were invented by ChatGPT — which he used to "enhance" a brief he then never re-read. Months later, a federal judge sanctioned three attorneys at a 350-lawyer firm, threw them off the case, and referred them to the state bar — and here is the part that matters most for assurance: the firm already kept a written AI policy. A policy on paper did not stop it. A public database that tracks these cases now counts more than four hundred eighty of them worldwide, and the line on the graph climbs every month. Read that as the consequence clock made concrete. Every one of those sanctions is a decision — to file, to sign, to vouch — that no one assured before it went out the door, and a cost that arrived afterward with a docket number. And the firm with the policy is the sharpest lesson of all: it proves that a rule people are told to follow is not the same as a control the process will not let them skip. The auditable, enforced verification step the coming profession exists to provide is exactly what a memo cannot supply. The bill for skipping it is no longer theoretical; it is being totaled, case by case, in the public record.

Draw the two curves and they cross in the same eighteen months: the cost of assuring a decision falling toward nothing, the number of decisions that need assurance climbing toward the sky. That crossing is the founding date of a profession.

Let me be precise, because precision is the whole point of this book. It is too strong to say no one audits decisions today. Your internal auditors, your risk committees, your model-validation teams, your own Architecture Review Board — every one of them assures some decisions, some of the time. What is missing is not the instinct; it is the discipline. No profession yet assures decisions completely, continuously, provably, and at scale — with a body of knowledge, a standard, a credential, and a seal. The function exists in fragments. The profession does not. That is what the rest of this chapter is for.

And notice the modest, defensible claim underneath all of this. We are not predicting that this profession will someday arrive. We are naming it, defining it, and dating its birth — which, as these pages will argue, is the surest way to be a founder rather than a footnote.

So meet the role the moment calls for: the person who does not make the decision, but assures it — who reads the rationale a stranger could not find, checks the evidence no one verified, and places, or withholds, a seal that says *this was decided well.* We kept circling this person for six chapters — the clipboard above the loop, the independent eye, the one who audits the auditor. It is time to give the job a name.

The auditor who makes nothing

The first thing to know about this person is what they do *not* do: they do not make the decision. That belongs to you, and it stays with you. The decision auditor makes nothing. Their entire craft is to take a decision someone else made and answer one question a stranger could trust — was this decided well, and can we prove it?

Picture the day. A decision arrives at their desk — a funding call, an architecture sign-off, a credit approval — and the auditor does not ask whether they would decide it the same way. That is not their job, and asking it would quietly turn them into a second decider, which helps no one. Instead they run the decision against the five properties. Was every criterion that should bear on this actually weighed — is it complete? Did the same standard that governed the last one govern this — is it consistent? Did the rules that must never break hold — is it governed? Does the reasoning rest on evidence a stranger could check — is it grounded? And is the whole of it — the verdict, the options, the reasons, the dissent — captured where someone could find it a year from now — is it on the record? They confirm the decision's level, they check that someone independent of the decider was involved where the level demands it, and at the gravest levels they review the futures the simulator flew. Then they do the one thing that makes them an auditor and not an advisor: they render a verdict on the *making* of the decision, and they sign their name to it.

If that sounds familiar, it should. It is exactly what a financial auditor does to a set of books — never writing the numbers, only assuring that the numbers were arrived at properly and can be believed by someone who was not in the room. We are simply pointing that ancient, trusted craft at the one thing it never covered: the decision itself.

A discipline, not a knack: the body of knowledge

A craft practiced by one gifted person is a knack. A craft anyone can learn, to a known standard, is a discipline — and only a discipline can become a profession. The difference between the two is a body of knowledge: a written, shared, teachable account of what the work is and how it is done well.

Every assurance profession rests on one. The internal auditors stand on their International Professional Practices Framework.

Project managers codify theirs; business architects codify theirs. None of these made the work easy, but each made it *transferable* — it could be taught, examined, and held to a standard, instead of dying with the one expert who happened to be good at it.

Decision Assurance needs the same, and this book is the first draft of it. The Decision Assurance Body of Knowledge gathers what the foregoing chapters built — the five properties, the levels, the two ledgers, the method of an audit, the standard a seal must meet, and the ethics the work demands. It is meant to be owned by no single company and stewarded in the open, by a council of practitioners, the way the disciplines we respect were built — by the people who do the work, for the people who do the work. That is not a sentimental preference. A body of knowledge that one vendor controls is a marketing brochure; a body of knowledge the practitioners own is a profession.

The credential: the Certified Decision Auditor

A body of knowledge makes the work teachable; a credential makes it trustable. When an organization lets someone place a seal on its decisions, it needs a way to know that person learned the discipline, works to its standard, and answers to a code beyond their own paycheck. That is what a credential is for. The Certified Public Accountant, the Certified Information Systems Auditor, the Certified Internal Auditor — each is a promise, backed by examination and ethics and continuing study, that the holder can be trusted with the word *assured*.

The Certified Decision Auditor will be the same kind of promise, and it will be earned the same slow way. Here I will be plain with you, the way this book stays plain throughout: a new credential is worth nothing on the day it is minted. The Certified Public Accountant took the better part of a century to mean what it means now. So this one will be built carefully and held to its standard without flinching, because the only thing a seal sells is trust, and trust arrives slowly and leaves fast. Better a small profession that means something than a large one that does not.

The engagement: the Decision Audit

What does the auditor actually deliver? An engagement, with a name worth saying plainly: the Decision Audit. And it audits a thing most people would not think to audit — not the outcome, and not even a single decision in isolation, but the *process* by which decisions are made.

That distinction matters enough to slow down on. Audit a single outcome and you are back to resulting — judging the decision by how it turned out. Audit the *process* and you are doing what every mature assurance profession does. A financial audit does not certify that a company will prosper; it certifies that its books were kept properly. A Decision Audit does not certify that a decision will turn out well; it certifies that the decision was *made* well — complete, consistent, governed, grounded, and on the record, to the level its consequence demands.

It comes in two sizes, and the level sets which one applies. For most of an organization's decisions, the auditor certifies the *process* — the standing way a team makes a class of decision, reviewed over a period and sealed, so that every decision flowing through that process inherits the assurance, the way a certified factory line vouches for every part it makes. Only for the gravest, one-way-door decisions — the Level A calls — does the auditor sit with a *single* decision and assure it on its own, the way a financial-statement opinion attends to one specific set of books.

And whatever its size, the Decision Audit returns two things, never one. First, a grade: assured to Level B, say, with the evidence behind it. Second — and this is the half organizations actually need — a gap list: the precise places where assurance is thin and more control is owed. *Sound here. Needs work there.* That second half is the whole assurance-and-control idea from earlier in this book, delivered as a morning's reading a leader can act on.

What a Decision Audit checks

A reader of an engineering bent will want the procedure on a single card, so here it is — the auditor's own checklist, and the criteria a Seal of Assurance is granted against. (The guard rides here too: it confirms the decision was made well; it does not make, or re-make, the decision.)

Level. Is the decision's level set correctly by its consequence and reversibility — and not talked down for convenience?

The five properties. Can each one be shown — complete, consistent, governed, grounded, on the record — to the depth the level demands?

Independence. Where the level requires it, was the check made by someone who does not answer to the decider?

The record. Could a stranger reconstruct the *why* a year from now, from what was written at the time?

The verdict. Assured to its level — or returned with a gap list naming exactly what to fix.

(This joins the Field Kit as the Decision Audit checklist.)

The mark: the Seal of Assurance

When an audit clears, it ends in a mark — and the mark carries a name we kept in reserve until now: the Seal of Assurance. You already trust marks like it without thinking. The auditor's signature on a financial statement. The seal a security firm places on a website it examined. Each says, in effect, *an independent professional looked, to a defined standard, and will stand behind what they found.*

The Seal of Assurance says exactly that about a decision, and it says it in two careful registers. It names the *level of assurance* —

borrowing the auditor's own honest vocabulary: reasonable assurance, not absolute, or the lighter *limited* assurance when that is all the evidence supports. And it names the *Decision Assurance Level* the process meets — assured to Level B, and so on.

Now the single most important sentence in this chapter, because everything turns on it. The Seal of Assurance certifies that a decision was *made well and can be proven so.* It does not, and must never, certify that the decision is *right.* The moment a seal promises a good outcome, it stops being an audit opinion and becomes a prophecy — and we spent this whole book showing why no one, machine or human, can sell you that. The seal is the most valuable thing the profession offers and the easiest thing in the world to corrupt, and the line that keeps it honest is the same line Gabor drew and the same humility the audit profession built its name on: we assure the making, never the future. A sound decision can still meet bad luck, and the seal stays true. That is not a weakness in the seal. It is the only thing that makes it worth trusting.

So who audits the auditor?

Which brings us, at last, to the line this chapter carried at its head — and to the obvious objection it invites. We just built a new authority and handed it a seal. So who watches them? If the decision auditor can be wrong, or captured, or lazy, what assures the assurer?

It is the right question, and the honest answer is that the profession does not escape its own discipline — it submits to it. The decision auditor is the most audited person in the building. Their independence is structural: they cannot audit a decision they helped make. Their work is itself on the record, so a peer can re-walk it. Their seals are reviewable, and a council of their peers holds the standard they are judged against. And the credential they carry can be lost. That is precisely how every assurance profession answers the same question about itself — the auditors are audited, by their standards, their independence, their peers, and their own visible record.

So the line is not a paradox to be escaped; it is a loop to be closed, the way the audit world closed it long ago. *Somebody must audit the auditor*. The answer is that the discipline does — and the day it stops, the seal is worth nothing, and we will say so first.

The two edges of the discipline

Before we leave the auditor, let me mark the two edges of what this discipline can do — one case showing what its absence costs, the other showing what even its presence cannot stop. Honesty about both is part of earning the seal.

Start with the absence. Theranos promised to run a full menu of blood tests on a single drop from a finger-stick, and on that promise it put its device into real clinics, returning results that patients and their doctors used to make medical decisions. Inside the company, the technology never worked as advertised — so, quietly, many of those tests were run on modified commercial machines instead. What never happened was the one thing that defines this entire profession: no credible, independent body ever validated the technology against its claims. And when employees who saw the failing results — among them a young scientist named Tyler Shultz — tried to raise the alarm, they were not heard and recorded; they were pressured, threatened, and pushed out before they could reach a regulator. Strip the case down to its assurance skeleton and you find a decision deployed with no grounding and no independent check, and the internal dissent suppressed rather than recorded. Independence was not defeated at Theranos. It was simply never let in the door — and that absence is exactly the void the decision auditor exists to fill.

Now the harder edge — the one I owe you, because a book that overclaimed would fail its own test. Volkswagen marks the outer limit of assurance, and it does so precisely because nothing about it was an honest mistake. Engineers wrote software that could detect when a car was sitting on an emissions test, switch the engine into a clean mode for the length of the test, and then pollute many times over the legal limit the moment the test was done. This was not a check that no one ran. It was a deliberate decision to defeat the check by design, built into roughly eleven million vehicles, and it held until outside researchers stumbled onto the discrepancy. The company pleaded guilty and paid billions. Sit with what Volkswagen does and does not say about assurance. It does not say the discipline is useless against fraud — independence, an immutable record, and a revocable seal all raise the cost of cheating and sharply raise the odds of being caught. But it says, plainly, the thing every honest assurance profession admits about itself: no seal stops a determined cheat. Assurance rests on a presumption of good faith; it makes a sound decision provable, and it makes a dishonest one far more dangerous to attempt — but it cannot read the heart of a person who already decided to lie. That is not a flaw to engineer away. It is the boundary of the claim, stated out loud — and stating it is itself an act of the discipline. The seal that promises to catch every fraud is the seal you should trust least.

So what?

For you, this is the difference between *hoping* your decisions are sound and being able to *prove* it to someone who was not there. It is the move from a culture where good judgment lives in a few trusted heads and leaves the day they do, to one where the quality of *deciding* is itself something you can measure, certify, and improve — like the quality of a product, or a financial statement, or a flight. The decision auditor is how that culture gets a backbone.

What now?

You will not hire a Certified Decision Auditor tomorrow; the credential is still being built. But you can do the thing the role does, this week, on one decision that matters. Take a recent Level A or B call, and audit it the way an outsider would: run the five-property checklist above, mark where the record goes thin, and write the one-line gap list. That makes it the first Decision Audit in your organization's history — and you will feel, immediately, how different *assuring* a decision is from *making* one.

Where we go next

We now hold the discipline, the role, the audit, and the seal. But a profession that judges decisions needs one more thing before it can judge them well across the years: a memory. How an organization records what it decided and why, and learns from how those decisions turn out — without letting a bad outcome rewrite a call that was sound when it was made — is the subject of the next chapter.

Chapter 7 in review

Key takeaways

- A new assurance profession is born when a technology turns **consequential and auditable at once** — the decision auditor is the next, on schedule after the financial and information-systems auditors.
- The convergence is real and datable: the cost of assuring a decision collapsed (≈280-fold in eighteen months) just as artificial intelligence made consequential decisions multiply and turn opaque — **AI is both the flood and the ark.**
- The decision auditor **makes nothing**; they assure the *making* of a decision against the five properties, at its level, with independence — and sign a verdict.

A profession needs four things: a **body of knowledge** (the Decision Assurance Body of Knowledge), a **credential** (the Certified Decision Auditor), an **engagement** (the Decision Audit), and a **mark** (the Seal of Assurance).

A **Decision Audit** assures the *process*, not the outcome (auditing an outcome is resulting), and returns a **grade plus a gap list** — "sound here, needs control there."

The **Seal of Assurance** certifies a decision was *made well and provably*, in the honest register of *reasonable, not absolute* assurance — **never that the decision is right.** That guard is the only thing that makes it worth trusting.

Somebody must audit the auditor is answered by the discipline itself: independence, peer review, a public record, and a credential that can be lost.

Key terms

Decision auditor — an independent professional who assures the making of a decision (against the five properties, at its level) without making the decision.

Decision Assurance Body of Knowledge — the shared, open, teachable account of the discipline; what turns a knack into a profession.

Certified Decision Auditor (CDA) — the credential: examined competence, a code of ethics, continuing study, and the standing to place a seal.

Decision Audit — the engagement that assures a decision *process* (or, at Level A, a single decision) and returns a grade plus a gap list.

Seal of Assurance — the mark stating the level of assurance (reasonable or limited) and the Decision Assurance Level a process meets; an audit opinion on the making, never a prophecy about the outcome.

Reasonable versus limited assurance — the auditor's honest registers of confidence: high-but-not-absolute, or the lighter form when the evidence is thinner.

Check your understanding

Why does a new assurance profession lag its technology by about a generation?

What is the one thing a decision auditor must *not* do, and why?

Why does a Decision Audit assure the *process* rather than the outcome?

What two things does a Decision Audit return?

What does the Seal of Assurance certify — and what must it never certify?

"Somebody must audit the auditor." In one line, what is the answer?

Answer key: (1) The world must first depend on the technology, and the tools to inspect it must mature; only then is it both consequential and auditable. (2) Make, or re-make, the decision — that turns the auditor into a second decider and destroys independence. (3) Auditing an outcome is resulting; the process is what can be assured, the way a financial audit certifies the books were kept properly, not that the company will prosper. (4) A grade (assured to a level) and a gap list (where more control is owed). (5) That the decision was made well and can be proven; never that it is right or will turn out well. (6) The discipline does — through independence, peer review, a public record, and a revocable credential.

Questions for reflection and discussion

Who in your organization already does fragments of this — internal audit, risk, model validation, the Architecture Review Board — and what would it take to make it a discipline rather than scattered instinct?

Where would an independent Seal of Assurance change behavior most — and where would it be resisted hardest?

For facilitators: ask the room who should hold the credential — a brand-new role, or your existing auditors and architects, retrained? The disagreement reveals how each organization will actually adopt this.

Apply it

Run a one-hour Decision Audit on a real, closed Level A or B decision from the past year — ideally one you did *not* make. Use the five-property checklist from this chapter: score each property to its level, test for independence, and try to reconstruct the *why* from the written record alone. Produce the two deliverables — a one-word grade and a gap list of no more than five items. *Trainers:* ask two people to audit the same decision separately, then compare gap lists; where they diverge is where your organization's assurance standard is still unwritten.

Notes and further reading

The professional precedents: the financial auditor and the Certified Public Accountant; the American Institute of Certified Public Accountants' Trust Services Criteria and the WebTrust seal; Service Organization Control (SOC) 2 reports; ISACA's Certified Information Systems Auditor; the Institute of Internal Auditors' International Professional Practices Framework and Certified Internal Auditor; and the International Standard on Assurance Engagements (ISAE) 3000, which defines *reasonable* and *limited* assurance for non-financial subject matter.

The why-now figures: Stanford Institute for Human-Centered Artificial Intelligence (HAI), AI Index Report 2025 (inference-cost collapse); Andreessen Horowitz, "LLMflation." The information-systems audit profession organized in 1969 (later ISACA). The regulatory clock: the European Union AI Act; the data-protection right to human intervention (GDPR Article 22); banking model-risk rules. The financial-audit profession hardened with the United States securities laws of 1933–34.

The "made well, not right" guard restates the soundness-versus-outcome distinction of Chapter 4 and the *reasonable, not absolute* humility of Chapter 5; the figure for the convergence is the chapter's "Why now" diagram.

CHAPTER 8

The Two Ledgers

> *Strategy is a pattern in a stream of decisions. — Henry Mintzberg, 1978*

Eight months after the decision, the email arrives.

You know the one. It lands on an ordinary afternoon, polite and quietly devastating: *Can you walk us through why we approved the vendor migration last fall?* The audit team is not accusing you of anything. They simply want the reasoning. And you realize, reading it twice, that the reasoning lives nowhere you can point to. It lived in your head, in a meeting no one minuted, in a deck that two reorganizations buried, in a chat thread that scrolled into the dark months ago. You made a good call. You remember that much. What you cannot do is *prove* it — not quickly, not cleanly, not in a way that satisfies a stranger who was never in the room.

That is the Introduction's nightmare, and I promised we would return to it. We are back. This chapter is the answer, and the answer is a peculiar one: the cure for that afternoon is not a better memory. It is a better *ledger*.

It helps to name the thing we are *not* building: an archive. An archive stores a record so it can be found later, and most of what an organization calls its "decision records" is exactly that — a document filed, retrievable in theory, opened by no one. An archive is passive; it asks nothing of the decision it holds. A ledger asks everything. It does not merely store the decision; it holds the decision to a standard — the reasoning, the evidence, the grade —

so the record does not just survive, it *stands up*. The distance between a record you filed and a record you can defend is the distance this whole chapter means to close.

Let me show you why a decision needs books — two sets of them — and why keeping them apart is the most disciplined thing your organization will ever do.

A decision dies twice

An unassured decision dies twice.

The first death is amnesia. The rationale evaporates. Within a quarter, the evidence you weighed, the options you set aside, the conditions you attached, the colleague who dissented and the reason you overruled the dissent — all of it thins into a vague recollection and then into nothing. The decision still stands; the *reasoning* is gone. So when the audit email comes, you reconstruct from fragments, and reconstruction always looks like rationalization, even when it is true. The wheel gets reinvented next year because no one remembers the last turn. You know this death well: it is the sound of the same debate opening for the third time, with none of the first two written down.

The second death is worse, and subtler. The organization remembers the *outcome* and judges the *decision* by it. The vendor migration went badly — a delay, an overrun, a public stumble — and now a room full of people who were never in the original room are certain it was an obvious blunder. The reasoning that made it a sound call at the time is invisible to them, because no one kept it. All they see is the wreckage, and the wreckage feels like proof.

Chapter 4 gave that second death a name: *resulting* — the habit, as Annie Duke puts it, of equating the quality of a decision with the quality of its outcome. (She borrowed the word from poker, where players learn the hard way that a perfect bet can lose and a reckless one can win.) Resulting is the most expensive memory error an organization makes, because it teaches exactly the wrong lessons. It punishes the people who decide well and draw a bad card. It rewards the people who decide badly and draw a good one. And it teaches everyone watching that the safe move is to optimize for outcomes they cannot control and to bury the reasoning that would expose how much luck was involved.

Two deaths, opposite in shape, identical at the root: the organization keeps the wrong ledger, or no ledger at all.

From what happened to what we decided

Here is the shift this chapter asks you to make, and it is a shift in what you choose to *remember*.

The instinct of every organization is to remember **what happened**. Outcomes are loud. They show up in the numbers, the post-mortems, the reputations that rise and fall in the hallway. Decisions are quiet. They happen in a moment, on the evidence then available, and then the moment passes and the outcome takes over the story.

Decision Assurance asks you to remember the quieter thing: **what we decided, and why, on what we knew at the time.** Not instead of the outcome — alongside it, and kept carefully separate from it. The paradigm is a move from outcome-memory to decision-memory, from *what happened* to *what we chose and on what grounds*.

If that sounds like extra bookkeeping, good — because bookkeeping is exactly the precedent. Five centuries ago, commerce faced a smaller version of this problem. You cannot run a business on a single number — how much cash sits in the drawer tonight — because that number tells you nothing about what you are owed, what you owe, or how you got here. So merchants learned to keep books, and in 1494 a Franciscan friar named Luca Pacioli wrote the method down: every transaction recorded twice, in two places, so the books could be checked against themselves.

He did not invent double-entry bookkeeping. Venetian traders were already using it. He simply gave it a printed form, and on that quiet discipline a great deal of the modern world was built.

A decision deserves books too. And like Pacioli's merchants, we keep them in two places on purpose.

The framework: the Assurance Record and the Outcome Ledger

Two ledgers. Name them, because naming is how a discipline remembers itself.

The Assurance Record is the *per-decision* ledger. It is the complete, sealed snapshot of a single decision at the instant it was made: the question on the table, the options considered, the evidence and where it was grounded, the assurance level the decision was held to, the conditions attached, the dissents recorded, and the name of the human who owned the call. It is immutable. Once sealed — and you will remember the Seal from Chapter 7 — it captures the decision *as it was*, on the evidence then in hand, not as it looks in the flattering or unforgiving light of hindsight. The Assurance Record is the answer to the audit email. When the question comes eight months later, you do not reconstruct. You retrieve.

The Outcome Ledger is the *cross-decision* ledger. It is what actually happened, tracked over time, decision after decision: did the migration land, did the risk materialize, did the bet pay. Outcomes live here — and they live here *deliberately*, in a separate book from the reasoning that preceded them.

Why two books and not one? Because the single most important rule in this entire chapter depends on keeping them apart.

The one rule: outcomes calibrate; they never re-grade

Read this rule slowly, because it is the hinge on which the whole framework turns.

The Outcome Ledger never reaches back and rewrites the Assurance Record. A decision graded well-made, on the evidence available at the time, stays well-made — even when the outcome is a disaster. Outcomes are allowed to *calibrate* the system going forward. Outcomes are never allowed to *re-grade* a decision after the fact.

This is the anti-resulting guard from Chapter 4, turned from a good intention into an architecture. The reason resulting is so hard to resist is that hindsight is overwhelming: once you know the migration failed, you cannot un-know it, and every reasonable-looking step that led there starts to look negligent. The only durable defense against that distortion is structural. You keep the grade and the outcome in separate books, and you write the rule in stone: a bad result may teach us, but it may not falsify the past. The decision was made in a moment that did not include the outcome. To judge it by the outcome is to judge it by information it could not possibly use.

A surgeon does everything right and loses the patient. An architect makes the sound call — the evidence supports it, the level is correct, the conditions are sound — and the market punishes it anyway. In a one-ledger world, both are blamed, because the only book is the book of outcomes. In a two-ledger world, the Assurance Record shows they decided well, the Outcome Ledger logs that it went badly, and the two facts sit side by side without contradiction. The organization learns the true lesson — *the world is uncertain; decide that way again* — instead of the false one: *that person is reckless; never approve that again.* Memory done right makes an organization braver and fairer in the same stroke.

What the Outcome Ledger is actually for

If outcomes never re-grade, you may wonder what they are good for. The answer is the most valuable thing memory does: it *calibrates*.

A single decision tells you almost nothing. Ten thousand decisions, each with its sealed reasoning and its recorded result, tell you a great deal — but only if you keep the stream where you can see it. Across that stream, patterns surface. Our Level-B calls in vendor risk turn out worse than their grade implied. Our grounding keeps missing the same class of evidence — the contract clause, the downstream dependency, the regulatory footnote. Decisions made under time pressure cluster, quietly, in the regret column. None of this re-grades any past decision. All of it sharpens the next one: it tells you where to raise a level, tighten a checklist, add a question to the gate, distrust your own confidence.

This is Mintzberg's insight, made operational. Strategy, he wrote, is not only the grand plan you deliberately set; it is also *a pattern in a stream of decisions* — the shape that emerges, often unintended, from the thousands of small calls an organization actually makes. The trouble is that most organizations never see the stream. The decisions scatter, the reasoning evaporates, and the pattern stays invisible until it shows up as a crisis. The Outcome Ledger is where you finally see the stream — and where the emergent pattern becomes something you can read on purpose, instead of something that reads you.

One decision, two books

Abstractions are easy to nod at and hard to keep. So let me walk a single decision through both ledgers, start to finish, and you will watch the whole machine turn.

Return to the vendor migration — the one the audit email asked about. Rewind to the afternoon it was decided, the previous spring.

The Assurance Record, sealed in March. The question on the table: replace the aging in-house order system, or buy the market-leading platform and migrate onto it. The record, written as the decision was made, holds the shape of the reasoning:

The question. Build a replacement in-house, or buy and migrate to the vendor platform, within an eighteen-month window.

The options weighed. Three — keep extending the legacy system; build new; buy and migrate. Each with its costed case, not only the winner.

The evidence, and where it was grounded. The vendor's reference customers, contacted directly. The internal build estimate, traced to the engineering leads who signed it. The legacy system's failure rate, pulled from the incident log, not from memory.

The level. Level B — a heavy door, costly to reverse, a large sunk commitment. Independent review required; the forward pass of the simulator run on the migration's key dependencies.

The conditions attached. Go-live gated on a data-integrity audit. A six-month exit clause negotiated into the contract, in case the platform underdelivered.

The dissent, recorded. One architect argued the legacy system could last three more years and the migration risk was underpriced. Overruled — and the reason for the overruling written down beside the dissent, not on top of it.

The owner. A name. One human, accountable for the call.

The expected outcome, pre-registered. This is the field everyone skips, and the one that makes the rest worth keeping: go-live within eighteen months; legacy retired; roughly $1.2 million saved against the build option over three years; user adoption above seventy percent within two quarters of launch.

Then the record is **sealed.** It captures the decision as it was, on the evidence then in hand. Nothing that happens next is allowed to reach back and edit it.

The Outcome Ledger, eighteen months on. The migration landed — but not the way the record predicted. Go-live slipped two quarters past the window, knocked sideways by a data-integrity problem the audit caught exactly where the conditions said to look. The savings came in real but smaller — closer to $800,000 than $1.2 million, once the overrun was counted. Adoption reached fifty-five percent, not seventy, and stalled there until a second training push moved it. The dissenting architect was partly right: the migration risk *was* underpriced, in precisely the dimension she named.

All of that goes in the Outcome Ledger. None of it goes in the Assurance Record.

Now the one rule does its work. A one-ledger organization — the kind with only the book of outcomes — reads this result and reaches one conclusion: *that was a bad call; the architect who dissented was right; do not trust that decider's estimates again.* It re-grades the past by the present, and it punishes a sound process for drawing a hard card.

The two-ledger organization does something else entirely. It opens the sealed Assurance Record and reads what was actually known in March. The evidence was grounded. The level was correct. The conditions were sound — and one of them *caught the very problem that caused the slip,* which is assurance working, not failing. The decision stays graded well-made, because it *was* well-made on what could be known. The grade does not move.

But the outcome is not wasted, either. It flows forward, into calibration. The delta between the pre-registered expectation and the recorded result becomes a lesson with a precise shape: *our migration estimates run optimistic on data-integrity effort, and our adoption forecasts run high by about fifteen points.* That lesson lands on the *next* vendor migration as a sharper question at the gate — *show me the data-integrity plan before I believe the timeline* — and a discounted adoption forecast. The dissent, preserved, is now worth its weight: the class of risk that architect flagged earns a standing line in the checklist. The institution decided well, drew a bad card, and got *wiser* — without rewriting its own history to do it.

And the audit email? When it arrives eight months after go-live asking why the migration was approved, the owner does not reconstruct anything. She retrieves the sealed record and forwards it: here is the question, here are the three options, here is the grounded evidence, here is the level and why, here is the dissent and why it was overruled, here is what we expected and how we are tracking against it. The reply that once cost a week of archaeology and ended in a shrug now costs four minutes and ends in a document. That is the difference between keeping a memory and keeping the books.

Your own base rates

Calibration gives you something most decision-makers never possess: your own reference class.

When you face a decision that feels novel, the most reliable question is rarely "how do I feel about this one?" It is "decisions *like* this one — how did they actually go for us?" Daniel Kahneman and Amos Tversky called this the *outside view*: stepping out of the seductive details of the case in front of you and looking instead at the base rate of similar cases. Bent Flyvbjerg turned it into a working method for planners — *reference-class forecasting*, sound enough that the American Planning Association formally endorsed it in 2005. The idea underneath is old statistical sense: the crowd of similar past cases predicts better than the story of the single present one.

But the outside view needs a reference class to work, and most organizations cannot build one, because they never kept the record. The Outcome Ledger *is* that reference class — your history, faithfully logged, decision by decision. It lets you replace "I think this will be fine" with "decisions shaped like this one succeeded for us six times in ten, and the four failures shared a feature this one also shows." That is not a machine predicting the future. It is your own institution, finally able to learn from itself. The memory makes the next forecast honest.

The Army built a humble version of this and named it the **after-action review (AAR)** — a structured debrief, developed in the mid-1970s and later institutionalized at the National Training Center after it opened in 1981, that asks four plain questions: what did we expect, what actually happened, why the difference, and what will we do differently. The genius of the AAR is what it refuses to do. It does not re-fight the engagement or hunt for someone to blame. It reviews the *process* so the next process improves — the body count stays in one column, the lessons go in another. David Garvin later carried the practice into business as a cornerstone of organizational learning. The two ledgers are the AAR made permanent: review the deciding, not just the result.

Memory is what makes assurance compound

Here is where assurance stops being a checkpoint and starts being an advantage.

One assured decision is worth something — it answers the audit, it protects the decider, it holds up under scrutiny. But a *memory* of assured decisions is worth something categorically larger. Ten thousand sealed Assurance Records, paired with an honest Outcome Ledger, are no longer a filing cabinet. They are a learning institution and an asset on the books. They are the thing a competitor cannot copy, because it is the distilled record of how your organization actually decides, where it reliably goes wrong, and how it gets better — written in your own history and ownable by no one else.

This is the quiet turn in the argument. Through most of this book, assurance is a discipline you adopt to keep a single decision defensible. Memory is what makes the discipline *compound*. Each sealed record makes the next decision a little better calibrated, a little better grounded, a little harder to fool — and the gap between an organization that keeps its books and one that does not widens with every quarter. The ledger is the moat. It is also, not coincidentally, the seed of something larger: a pattern emerging from a stream, which is the shape of intelligence itself. We will return to that idea at the end of the book. For now it is enough to say that an organization which remembers how it decides is an organization that gets wiser on a schedule, and most do not.

The honest caveats

Memory carries its own failure modes, and a chapter that skipped them would betray the discipline it is selling.

A ledger no one reads is theater. The record only assures if someone retrieves it — at the next audit, the next similar decision, the next calibration review. Build the books to be opened, not merely filled.

Calibration on a thin or biased record misleads. Your reference class is only as honest as the outcomes you bothered to log, and the temptation is to record the wins and let the failures quietly fall off the page. A calibration built on a flattering ledger is worse than none, because it launders luck into apparent skill.

And the temptation to re-grade is constant. Every time an outcome lands badly, someone will want to reach into the Assurance Record and mark the decision wrong, to make the past match the present pain. Refusing that is the whole discipline. The bad outcome belongs in the Outcome Ledger, where it can teach. It does not belong in the Assurance Record, where it can only lie.

Keep the two books. Keep them apart. Let outcomes calibrate, and never let them re-grade. Do that, and the email that arrives eight months after the decision stops being a nightmare and becomes the easiest reply you will write all week — because the reasoning is not in your head, or a buried deck, or the dark of a scrolled-away thread. It is in the record, sealed on the day you decided, waiting for exactly this question.

So what?

An organization that remembers only outcomes is condemned to misjudge its own decisions — punishing the unlucky, rewarding the fortunate, and forgetting the reasoning the moment the audit needs it. Two ledgers, kept apart, end both deaths at once.

What now?

Pick one consequential decision your team made in the last year. Ask one question of it: *if an auditor asked us today why we decided that, could we retrieve the reasoning, or would we reconstruct it?* The answer tells you whether you keep an Assurance Record or merely a memory.

Where we go next

Memory tells you what your decisions are connected to *in time* — before and after. The next chapter takes up connection in a second dimension: how decisions depend on one another, and how a model of those dependencies lets you rehearse a future before you commit to it.

The chapter, in review

Key takeaways

- An unassured decision dies twice: once to **amnesia** (the reasoning evaporates and cannot be retrieved) and once to **resulting** (the decision is judged by its outcome rather than its process).
- The cure is not a better memory but a better **ledger** — and, like double-entry bookkeeping, two of them kept deliberately apart.
- The **Assurance Record** is the immutable, sealed, per-decision snapshot of the reasoning at the time it was made — the answer to the audit email.
- The **Outcome Ledger** is the cross-decision record of what actually happened, kept in a separate book from the reasoning.
- **The one rule:** outcomes calibrate; they never re-grade. A decision well-made on the evidence available stays well-made, even when the result is bad.
- The Outcome Ledger's job is **calibration** — surfacing the pattern in the stream (Mintzberg) and giving you your own **reference class** (the outside view) so the next decision improves without thc past being falsified.
- Memory is what makes assurance **compound**: a record of ten thousand assured decisions is a learning institution and an asset no competitor can copy.
- Memory fails too — an unread ledger is theater, a flattering record launders luck, and the urge to re-grade after a bad outcome is constant. The discipline is refusing it.

Key terms

The Assurance Record — the immutable, sealed, per-decision ledger: the complete reasoning, evidence, level, conditions, and ownership of one decision, captured as it was at the moment of decision.

The Outcome Ledger — the cross-decision ledger of what actually happened over time; the source of calibration, kept separate from the Assurance Record by design.

Calibrate, never re-grade — the governing rule that lets outcomes sharpen future decisions while forbidding them from rewriting the grade of a past one; the structural cure for resulting.

Reference class (the outside view) — the set of similar past decisions whose base rate predicts the present one better than its own details do; the Outcome Ledger is what makes your own reference class possible.

Check your understanding

Name the two ways an unassured decision "dies," and explain why they share the same root cause.

What distinguishes the Assurance Record from the Outcome Ledger, and why are they kept in separate books?

State the one rule in your own words. What failure does it exist to prevent?

If outcomes never re-grade a past decision, what *are* they allowed to do?

Why is a calibration built on a flattering (wins-only) Outcome Ledger "worse than none"?

Answer key. (1) Amnesia — the reasoning evaporates and cannot be retrieved; and resulting — the decision is judged by its outcome. Both come from keeping the wrong ledger, or none: remembering outcomes while forgetting reasoning. (2) The Assurance Record is per-decision, immutable, and captures the reasoning at the time; the Outcome Ledger is cross-decision and records what happened. They are separated so a later outcome cannot distort the recorded grade. (3) Outcomes calibrate; they never re-grade — it prevents resulting, the judging of a sound decision by an unlucky result. (4) Calibrate the system going forward: surface patterns, adjust levels and checklists, and build the organization's reference class. (5) Because it launders luck into apparent skill — a reference class drawn only from wins tells you the base rate of success is higher than it truly is, and miscalibrates every future decision built on it.

Questions for reflection and discussion

Think of a decision in your organization that was judged by its outcome rather than its process. What would the conversation about it sound like if an Assurance Record existed?

Where does your team currently store the *reasoning* behind consequential decisions — and could you retrieve it eight months later?

For facilitators: Ask the group to surface a decision that "turned out badly." Then ask whether anyone can reconstruct the evidence available at the time. Use the gap between the confident judgment and the thin reconstruction to make resulting visible.

What would it take, culturally, for your organization to log its failures into the Outcome Ledger as faithfully as its wins?

Apply it

Choose one decision your team made in the past year and build a minimal Assurance Record for it after the fact: the question, the options considered, the evidence and its source, the level of consequence, the conditions, and the owner. Then, in a separate note, record the outcome. Notice how unnatural it feels to keep them apart — and how much clearer the decision looks once you do. *Trainers:* run this with two subgroups on the same decision;

compare the records they reconstruct. Where they differ is precisely the reasoning your organization never wrote down — and the strongest argument in the room for keeping the Assurance Record in the first place.

Notes and further reading

Resulting / outcome bias. Annie Duke, *Thinking in Bets* (Portfolio, 2018) — the term as used here, borrowed from poker and popularized for decision-making.

Strategy as a pattern in a stream of decisions. Henry Mintzberg, "Patterns in Strategy Formation," *Management Science* 24, no. 9 (1978); and Henry Mintzberg & James A. Waters, "Of Strategies, Deliberate and Emergent," *Strategic Management Journal* 6, no. 3 (1985).

The outside view and reference-class forecasting. Daniel Kahneman & Amos Tversky, "Intuitive Prediction" (1979); Daniel Kahneman, *Thinking, Fast and Slow* (2011), ch. 23; Bent Flyvbjerg, "From Nobel Prize to Project Management" (2006); reference-class forecasting endorsed by the American Planning Association in 2005.

Forecaster calibration. Philip E. Tetlock, *Expert Political Judgment* (2005) and *Superforecasting* (with Dan Gardner, 2015); the Brier score (Glenn W. Brier, 1950) measures the overall accuracy of probabilistic forecasts, decomposable into calibration and resolution.

The after-action review. Developed by the U.S. Army in the mid-1970s and institutionalized at the National Training Center (opened 1981); carried into business by David A. Garvin, *Learning in Action* (Harvard Business School Press, 2000).

Double-entry bookkeeping. Luca Pacioli, *Summa de Arithmetica* (Venice, 1494) — the first printed description of a method merchants already used; the earlier manuscript of Benedetto Cotrugli (1458) went unprinted until 1573.

The remembered past. "Those who cannot remember the past are condemned to repeat it" — George Santayana, *The Life of Reason*, vol. 1, *Reason in Common Sense* (1905).

PART IV

The Tool and the Practice

CHAPTER 9

The Flight Simulator

The future cannot be predicted, but futures can be invented. — Dennis Gabor, 1963

The engine fails four seconds after takeoff.

The cockpit lurches, the alarms scream, and the pilot's hands move before the fear does — throttle, rudder, the climb traded for airspeed, the field already chosen. The aircraft steadies. Then a door opens in the side of the world, a technician leans in, and says: *nice work — let's run it again.* None of it was real. The runway was a screen. The lurch was hydraulics. The emergency was a line of code. The pilot just rehearsed, for the fortieth time, a failure they will most likely never meet — so that if they ever do, the hands already know the motions.

This is one of the quietly radical inventions in the history of flight, and it is older than the jet engine. In 1929 a young organ-builder named Edwin Link bolted a cockpit onto bellows and valves borrowed from player pianos and built the first widely used flight simulator — the "Blue Box." For a few years the aviation world treated it as a fairground toy. Then, in 1934, the U.S. Army Air Corps took over carrying the mail, flew into bad weather it was not trained for, and lost a string of young pilots in the span of weeks. The Army bought six of Link's boxes that same year. By the end of the Second World War, more than half a million airmen learned to fly on instruments inside them — rehearsing, on the ground, the emergencies that would otherwise teach their lessons only once, and at terrible cost.

Now hold that image against the way you make your most consequential decisions. You make the call once. Live. With everything riding on it. And you discover what it does to everything around it only after it is done and cannot be taken back. The pilot rehearses the engine failure forty times in safety; you meet yours in flight, at altitude, with passengers aboard. This chapter is about closing that gap — about building, for decisions, the thing aviation built for pilots a century ago. To do it, we need two pieces: a way to *see* how decisions connect, and a way to *fly* them before we commit. The map, then the simulator.

Decisions are connected, and the connection is invisible

The last chapter showed decisions connected in *time* — each one before and after the others, remembered in two ledgers. This chapter shows them connected in *structure*, and structure is where the danger hides.

No decision stands alone. Every consequential call rests on decisions and assumptions that came before it, and supports decisions that will lean on it later. Your vendor migration rested on a roadmap promise, a budget assumption, and an earlier platform standard. Three later decisions — an integration design, a staffing plan, a compliance sign-off — quietly rested on the migration. Pull any one node out of that web, or let one assumption rot, and the others wobble. The roadmap promise breaks; and somewhere downstream, three decisions that depended on it are now unsafe, and no one knows, because no one can see the thread that connects them. They fail later, one at a time, looking like bad luck. They were not bad luck. They were an unmapped dependency.

This is the everyday tragedy of decisions made in isolation. The organization treats each call as a self-contained event, files it (if you are lucky), and moves on — blind to the web that makes a small broken assumption over here detonate a large failure over there. We would never run a supply chain this way, or a building's load-bearing structure, or a flight plan. We run our decisions this way constantly.

The Decision Dependency Graph

So make the web visible. Name it: **the Decision Dependency Graph.**

The idea is simpler than it sounds, because you already think in graphs without calling them that. A graph is just nodes connected by lines. Here the nodes are decisions, and the lines are dependencies — directed lines, pointing from a decision to the decisions and assumptions it rests on. Read the arrows one way and you see a decision's *antecedents*: everything it stands on. Read them the other way and you see its *dependents*: everything standing on it. The whole tangled web of "this rests on that" becomes a structure you can hold in your hand, query, and trace.

None of this is exotic. Engineers manage dependency graphs every day — a software build will not compile until the things it depends on are built first, in the right order. Closer to your world, enterprise architecture lives and breathes dependency maps: which application feeds which, which capability rests on which platform. And the discipline of project management mapped dependencies into a network seventy years ago — the **critical path method (CPM)**, built by James Kelley and Morgan Walker in 1957, and the **program evaluation and review technique (PERT)**, built for the U.S. Navy's Polaris program in 1958, both turned a project's tangle of "this task waits on that task" into a graph you could see and schedule. We mapped the dependencies between *tasks* three generations ago. We simply never mapped the dependencies between *decisions*. The Decision Dependency Graph is that overdue map.

What the map alone buys you

Before you run a single simulation, the map itself pays for the effort — in three reads.

The first read is **blast radius.** Point at any decision and the graph shows you everything downstream that depends on it. Before you reverse a call, change a standard, or retire a system, you see exactly what you are about to shake.

Consider what a single unconverted number did to a spacecraft. In December 1998, NASA launched the Mars Climate Orbiter — a robotic explorer, the product of years of work and roughly 125 million dollars. For nine months and 416 million miles, it flew flawlessly. The ground software that computed the thruster firings, written by one contractor, reported its results in pound-force seconds, the imperial unit. The navigation software that consumed those results, written by another team, assumed newton-seconds, the metric unit. The two numbers differed by a factor of about four and a half. No one checked the seam where they met.

The error did not announce itself. It accumulated — quietly, across every course correction of the cruise — bending the spacecraft's true path a little further from its modeled path with each passing week. When the orbiter finally reached Mars on September 23, 1999, and fired its engine to slip into orbit, it came in roughly a hundred kilometers too low, into atmosphere too thick to survive. It went silent behind the planet and never came back out. A 125-million-dollar machine, lost to a unit of measurement nobody wrote down.

That is what blast radius looks like when no one can see it. One ungrounded assumption flowed unnoticed into everything downstream, and the failure surfaced only at the end, when the cost of finding it was total. The graph is the thing that draws that thread in red before the burn, not after the silence — that points at the single ungrounded node and shows every decision standing on it, while there is still time to check the seam.

The second read is **provenance** — the answer to *why*. Trace a decision backward to the antecedents and assumptions it rests on, so that the day an assumption breaks, you can find every decision that quietly stood on it and check whether each still holds. The broken roadmap promise stops being a delayed surprise and becomes an immediate, visible alert on three named decisions.

The third read is the **single point of failure** — the assumption that a dozen decisions secretly share. On a graph, that lonely, overloaded node is obvious; in a filing cabinet, it is invisible until the day it gives way and a dozen decisions fall together. Seeing it early is the difference between reinforcing a beam and explaining a collapse.

From map to twin

A static map is valuable. A living one changes the game.

Keep the graph in sync with reality — the decisions, their assumptions, their current status, what is still true and what just quietly stopped being true — and the map becomes a **digital twin** of your decision system: a virtual mirror of the real thing, accurate enough to experiment on. The concept comes from engineering, where Michael Grieves described it around 2002 and NASA later gave it the name "digital twin" (the oldest ancestor is the set of paired ground models NASA flew alongside Apollo, the duplicates that let Houston rehearse a fix for Apollo 13 on the ground before asking the crew to try it in space).

That Apollo image is the whole point. The twin is a safe copy of reality where you can break things without breaking anything — a place to crash with no one aboard. Which is exactly what a flight simulator is. We now own the simulator's world. The last step is to fly it.

The Decision Flight Simulator

Name the capability for what it is: **the Decision Flight Simulator.** It runs *what-if* on the twin, and there are three ways to fly it.

Forward — what does this do? Propose a decision and let the simulator push its consequences across the graph. It traces the blast radius, names the assumptions the decision would stress, and flags the dependents it would put at risk. You rehearse the move before you make it, and you walk the wreckage before there is any. The integration that three later decisions relied on shows up as endangered *before* you sign, not after it fails.

Outcome — how might it go? Run the decision under uncertainty, not once but many times, and watch the spread of ways it plays out. This is not one confident prediction; it is a cloud of rehearsals. The trick is old: when a problem is too tangled to solve directly, you stop trying to compute the answer and instead sample it, over and over, until the shape emerges. That is the **Monte Carlo method**, invented at Los Alamos in the 1940s by Stanislaw Ulam and John von Neumann for problems no equation could crack. The Decision Flight Simulator borrows their move to show you not *the* outcome but the *distribution* of outcomes — the range you are actually signing up for.

Backward — how do we get there? Name a future you want, and let the simulator work backward to the decisions that would produce it. The planners call this **backcasting** — a term John Robinson coined in 1982 — and it inverts the usual question. Instead of "where do our current decisions lead?" it asks "what decisions lead where we want to go?" You start at the destination and trace the route home.

One decision, flown

Let me fly one, so the three modes stop being abstractions.

Take the vendor migration from the last chapter — the one that slipped two quarters on a data-integrity problem. Rewind again to the spring it was decided, but this time, before anyone signs, you put it through the simulator. Here is what each of the three modes shows.

Forward — what does this do? You drop the proposed decision onto the graph and let its consequences ripple. The simulator traces the blast radius and lights up what the migration touches: the integration design that assumes the new platform's data model, the staffing plan built around the eighteen-month window, the compliance sign-off that rests on a clean data-integrity story. One node flashes the loudest warning — the data-migration step, where the legacy system's messy records meet the new platform's strict schema. The simulator cannot tell you the migration will slip. It can tell you that *if* anything breaks, this is the seam where it breaks, and three named decisions are standing on it. That is the emergency, flagged on the ground, months before the flight.

Outcome — how might it go? Now you run it under uncertainty — not once, but ten thousand times, each run sampling a different combination of how long the data cleanup takes, how fast users adopt, how the integration behaves. The simulator does not print a date. It prints a *spread*. In most runs the migration lands inside the window; in a meaningful tail — the runs where the data-integrity work drags — it slips one to two quarters. You came in expecting a confident "eighteen months." You leave knowing the truth: eighteen months is the *middle* of a range whose ugly tail is exactly the data-integrity risk the forward pass already flagged. The number you were about to commit to was the optimistic half of a distribution you could not see.

Backward — how do we get there? Finally, you name the future you want — *go-live in eighteen months, no compliance surprise* — and let the simulator work back to the decisions that reach it. The route it traces is blunt: the data-integrity audit cannot wait until just before go-live, where the original plan put it; it must move to the front, gating the contract, not the launch. Backcasting from the destination rewrites the sequence. The condition that, in the last chapter, *caught* the problem late now moves early enough to *prevent* the slip.

Notice what just happened. The simulator did not predict the future. It rehearsed three of them — the consequence map, the spread of outcomes, the route to the one you want — and every one pointed at the same node: data integrity. The emergency that, in the real timeline, surfaced eighteen months later as a two-quarter slip was sitting in plain sight in the rehearsal, before a signature, while it still cost nothing to move the audit forward.

That is the whole promise of the chapter, made concrete on a single decision. In the last chapter, the migration's conditions caught the problem — assurance working after the fact, exactly as designed. Here, the rehearsal sees the same problem *before the fact,* and hands you the chance to change the plan rather than merely survive it. Same decision, same risk, two different relationships to time. The ledger proves you decided well. The simulator gives you a fair shot at deciding *better*.

We do not predict the future. We rehearse it.

Here is the line that holds the whole capability honest, and it is the same humility that runs through every chapter of this book.

A flight simulator does not tell the pilot what will happen on Tuesday's flight. It cannot. What it does is let the pilot rehearse Tuesday's emergencies so thoroughly that, whatever actually happens, the hands already know the motions. The Decision Flight Simulator makes the same modest, powerful claim. It does not predict your future. It lets you rehearse it — the blast radius, the spread of outcomes, the route to the future you want — so that when you commit for real, you commit with the emergencies already flown.

Dennis Gabor said it in 1963, and it became this book's signature: the future cannot be predicted, but futures can be invented. Eisenhower said the working half of it to a roomful of planners in 1957, quoting an old Army saying: *plans are worthless, but planning is everything.* The plan you carry into the room may be wrong by lunch. The planning — the rehearsal, the hours in the box — is what lets you respond when it is. The simulator's value is not the forecast it prints. It is the readiness it builds.

So the outputs *inform*; they never *decide*. Reasonable, never absolute. The human still flies the plane. A decision simulator that claimed to know Tuesday would be exactly the oracle this book refused back in Chapter 3 — the smarter machine that quietly takes the controls. This one does the opposite. It rehearses, and it hands the controls back, and it leaves the deciding where it belongs: with the person who answers for it.

The precedent that already paid off

If this sounds aspirational, know that a manual version of it already saved a company, and the story is famous for a reason.

In the early 1970s, two strategists at Royal Dutch/Shell, Pierre Wack and Ted Newland, refused to hand their executives a single forecast. Instead they built scenarios — rehearsals of several futures, including one in which the price of oil shocked the world. They did not predict the 1973 embargo. They *rehearsed* it. So when it came, Shell's leaders already flew that emergency in their minds, and the company moved through the crisis from roughly the seventh-largest oil major to the second. The roots of their method ran back to Herman Kahn's scenario work at the RAND Corporation in the 1950s and 1960s — rehearsing unthinkable futures so that meeting one would not mean meeting it unprepared.

Scenario planning proved the principle with a once-a-year offsite and a handful of the boardroom's biggest questions. The

Decision Flight Simulator is that same principle made continuous, connected, and computable — every consequential decision rehearsed, not just the annual three; on the live graph, not a binder that yellows on a shelf. Shell flew one emergency by hand and it was nearly enough. Imagine flying all of them.

Why this sits at the top of the value

It is worth saying plainly where this capability lands in the discipline, because it is the one executives pay the most to obtain.

Everything else in this book makes a decision *defensible*. Capture it, ground it, grade it, seal it, remember it — and you can prove, afterward, that the decision was sound. The dependency graph and the simulator add the one thing the others cannot: they make a decision *rehearsable*. They move you from knowing a decision was sound *after the fact* to seeing what it will do *before you commit*. That is the difference between an audit and a flight simulator — between the discipline that documents the outcome and the discipline that gives you a fair chance to change it. Not change it the way an oracle would, by knowing the future. Change it the way a rehearsed pilot does — by meeting the emergency already knowing the motions, by catching the endangered dependent while there is still time to shore it up.

That is the top of the value, and it is why this is the chapter the marquee belongs to.

The honest caveats

A flight simulator earns trust only by being honest about its limits, and so must this one.

A simulation is only as good as the graph beneath it. A wrong dependency or a stale assumption produces a confident, *wrong* rehearsal — and a confident wrong rehearsal is more dangerous than none, because it feels like knowledge. The twin must be maintained, or it lies with a straight face.

The crisp number seduces. A simulator prints a clean distribution that *looks* like certainty, and the human temptation is to treat the rehearsal as a prophecy. Resist it, or you rebuild the oracle you were right to reject. The output is a rehearsal. Read it as one.

The map is not the territory. The graph models the dependencies you thought to encode — never the ones you missed. The dependency that bites is almost always the unmodeled one, so the simulator makes you safer without ever making you safe, and the gap between those two words is where judgment lives.

And the twin costs real work to build and keep current. Which is why the Decision Assurance Levels from Chapter 6 matter here: they tell you which decisions earn the full flight rehearsal and which can fly on instinct. You do not simulate the lunch order. You simulate the migration, the reorganization, the irreversible bet — the decisions whose blast radius is large and whose door, once walked through, does not open the other way.

Build the map. Keep it living. Fly the emergency before it is real. Then make the call with your hands already knowing the motions — and hand the controls to no one.

So what?

Decisions fail in ways you cannot see because they depend on one another invisibly; a Decision Dependency Graph makes the web visible, and a Decision Flight Simulator lets you rehearse a high-stakes decision — its blast radius, its range of outcomes, its route to the future you want — before you are committed to it.

What now?

Take one consequential decision on your desk and draw its first-order dependencies by hand: what it rests on (antecedents) and what will rest on it (dependents). Even that rough sketch will surface a dependency you were carrying without seeing.

Where we go next

The map and the simulator are the most visible of the jobs an assurance tool must do. The next chapter assembles all of them — capture, grounding, the graph, the levels, the simulator, the seal, the memory — into a single picture of the tool that catches what we miss.

The chapter, in review

Key takeaways

Decisions fail invisibly because they **depend on one another** — every decision rests on antecedents and supports dependents, and a broken assumption detonates downstream where no one is looking.

The **Decision Dependency Graph** makes that web visible: decisions as nodes, dependencies as directed lines, queryable and traceable — the overdue cousin of the critical-path networks project managers built in the 1950s.

The map alone buys three reads: **blast radius** (what a change shakes), **provenance** (what a decision rests on), and the **single point of failure** (the assumption a dozen decisions secretly share).

Kept in sync with reality, the map becomes a **digital twin** — a safe copy of your decision system you can experiment on without consequence.

The **Decision Flight Simulator** runs what-if on the twin three ways: **forward** (what does this do?), **outcome** (how might it go, across many runs?), and **backward** (what decisions reach the future we want?).

The governing promise is humble and exact: **we do not predict the future, we rehearse it.** Outputs inform; they never decide. Reasonable, never absolute. The human still flies the plane.

Scenario planning proved the principle — **Shell rehearsed the 1973 oil shock** and moved from seventh to second — and the simulator makes that once-a-year discipline continuous, connected, and computable.

It earns its limits in the open: a simulation is only as good as its graph, a crisp number seduces, the unmodeled dependency is the one that bites, and the **Decision Assurance Levels** decide which calls deserve the full rehearsal.

Key terms

Decision Dependency Graph — a model of decisions as nodes connected by directed dependencies; reading the arrows reveals each decision's antecedents (what it rests on) and dependents (what rests on it).

Digital twin — a virtual mirror of the real decision system, kept in sync with it, on which experiments can be run safely; the simulator's world.

Decision Flight Simulator — the what-if capability run on the twin, in three modes (forward / outcome / backward); it rehearses a decision's consequences rather than predicting them.

Rehearse, do not predict — the discipline's stance on foresight: the value is the readiness the rehearsal builds, not a forecast it claims; outputs inform the human, who still decides.

Check your understanding

Explain why a decision made in isolation is dangerous, using the idea of antecedents and dependents.

Name the three reads the dependency map gives you *before* any simulation, and what each one shows.

What turns a static dependency map into a digital twin, and why does that matter?

Describe the three modes of the Decision Flight Simulator and the question each one answers.

The chapter insists the simulator "rehearses, not predicts." Why is that distinction the difference between this tool and the "oracle" rejected in Chapter 3?

Answer key. (1) Because every decision rests on antecedents (earlier decisions and assumptions) and supports dependents; in isolation, a broken antecedent silently makes its dependents unsafe, and the failures surface later looking like bad luck. (2) Blast radius — what a change to a decision shakes downstream; provenance — what a decision rests on, so a broken assumption can be traced to every dependent; single point of failure — the shared assumption many decisions secretly depend on. (3) Keeping the map in sync with reality (decisions, assumptions, status) turns it into a twin; it matters because you can then experiment — crash — safely, without touching the real system. (4) Forward: propose a decision and trace its consequences across the graph (what does this do?); outcome: run it under uncertainty many times to see the spread (how might it go?); backward: start from a desired future and work back to the decisions that reach it (how do we get there?). (5) Because predicting would put the machine in the controls (the oracle), while rehearsing keeps the human flying — the simulator informs readiness and hands the controls back, never claiming to know the future.

Questions for reflection and discussion

Think of a failure in your organization that "looked like bad luck." Trace it backward — was it really an unmapped dependency on a broken assumption?

Which decisions in your portfolio carry a blast radius large enough to deserve a full flight rehearsal, and which can safely fly on instinct?

For facilitators: Ask the group to pick one real decision and collectively draw its dependency graph on a whiteboard. Watch for the moment someone names a dependent no one connected before — that moment is the chapter's whole argument in miniature.

Where in your organization does a confident forecast already get treated as a prophecy — and what would change if it were read as a rehearsal instead?

Apply it

Choose one high-stakes decision currently in flight. Sketch its dependency graph two layers deep — its antecedents and their antecedents, its dependents and their dependents. Then run a manual forward simulation: pick one antecedent assumption, suppose it breaks, and trace by hand which downstream decisions become unsafe. Write down what you find. *Trainers:* give two subgroups the same decision and ask them to build the graph independently, then overlay the two; the dependencies only one group caught are the ones your organization is currently carrying blind, and the strongest case for building the map for real.

Notes and further reading

Invented futures. Dennis Gabor, *Inventing the Future* (Secker & Warburg, 1963): "The future cannot be predicted, but futures can be invented."

The flight simulator. Edwin Link's "Blue Box" (patented 1931); adopted by the U.S. Army Air Corps in 1934 after a run of air-mail crashes; more than 500,000 airmen trained on instruments during the Second World War — the first widely used flight simulator.

Planning over plans. Dwight D. Eisenhower, remarks to the National Defense Executive Reserve Conference, November 14, 1957: "Plans are worthless, but planning is everything" — quoted by Eisenhower as a longstanding Army maxim.

Dependency networks. The critical path method (James Kelley & Morgan Walker, 1957) and the program evaluation and review technique (PERT, U.S. Navy Polaris program, 1958) — the project-scheduling ancestors of the decision graph.

The digital twin. Concept articulated by Michael Grieves (≈2002, product lifecycle management); the term "digital twin" credited to John Vickers at NASA (≈2010); conceptual ancestor in NASA's paired Apollo-era models.

Sampling the answer. The Monte Carlo method — Stanislaw Ulam and John von Neumann, Los Alamos, 1940s; named by Nicholas Metropolis — for problems too tangled to solve directly.

Working backward. "Backcasting" — John B. Robinson, "Energy Backcasting," *Energy Policy* (1982): start from a desired future and trace the decisions that reach it.

Scenario planning. Pierre Wack and Ted Newland at Royal Dutch/Shell (early 1970s), credited with Shell's readiness for the 1973 oil shock; rooted in Herman Kahn's scenario work at the RAND Corporation.

CHAPTER 10

The Thing That Catches It

In that cafeteria, I made a promise to build the thing that would catch it. We spent the previous chapters earning the right to describe that thing — the five properties, the levels, the auditor and the seal, the two ledgers, the dependency graph and its simulator. Now let me describe the machine itself: not a product you buy on a particular Tuesday, but the *ideal tool* for Decision Assurance — what any system must do to deserve the name, and an honest reckoning of how close we can build it today.

One warning before we start, because it is the warning this whole book carries. The tool is not the discipline. Chapter 7 gave you the discipline — a human auditor, above the loop, who assures and does not decide. A tool cannot replace that person; the experiment in Chapter 3 settled it. What a tool *can* do is give that person reach: let one auditor assure a thousand decisions a week instead of three, by taking on the parts that are throughput and synthesis and leaving the judgment where it belongs. So read what follows as a description of the auditor's instrument, never the auditor's replacement.

What the machine must do

Strip away the dashboards and the demos and a Decision Assurance system does seven things. Each maps to something you already know; together they are the engine.

1. Catch the decision as it happens

Nothing else works without this one. A decision is the most perishable thing an enterprise produces — alive in the room, gone by the elevator. So the machine's first job is to **capture** the decision while it is warm: the verdict, the options weighed, the reasons, the evidence, the dissent — recorded as a by-product of deciding, not begged for in a meeting six months later. Remember the audit email from the very first pages of this book, the one you could not answer? The machine exists so that note is never unanswerable again. If a system makes recording a decision feel like extra work, people will not do it, and the discipline dies at step one. Capture must be nearly free.

2. Ground every claim

The machine must hold a single, unbending rule: every claim cites checkable evidence, or the claim is withheld. A vendor "chosen on a strong feeling," an estimate that traces to no model, a citation that resolves to nothing — the system flags each as ungrounded before it can travel into a decision. This is the lawyer's fabricated cases caught before filing; it is the London Whale's spreadsheet model grounded before a single trade rides on it. And there is a second, sharper edge: the machine's *own* outputs must be grounded too, because the model that assures can hallucinate exactly as fluently as the model that decides. *Who assures the assurer?* — the answer, inside the tool, is that the tool shows its evidence for everything it asserts, and abstains when it cannot. New York City learned the cost of the alternative in public: its official "MyCity" business chatbot, launched to help entrepreneurs, confidently told them they could take a cut of their workers' tips, turn away tenants who pay with housing vouchers, and refuse to accept cash — each answer fluent, helpful, and flatly against New York law. The bot cited nothing; it simply asserted, in the city's own voice, on questions that carry real legal consequences. A grounding rule —

point to the controlling statute, or do not answer — turns that confident wrongness into a blank space where a bad answer used to be.

3. See the whole — the dependency graph

Most of the failures in this book share one sentence: *no one was watching the whole.* The unverified unit at the seam between two teams. The security patch ordered and never confirmed applied. The deployment that left one server on old code. No human committee holds an enterprise's thousands of interdependent decisions in mind at once — but a machine can. The engine builds the **Decision Dependency Graph** of Chapter 9: what this one rests on, what it will bind downstream, which standing constraints it must not break, and where it quietly contradicts a decision made last quarter. This is completeness and governance rendered at a scale no person can reach by hand — the machine watching the whole that no single human can.

4. Grade the decision — the level

The machine must refuse to assure everything to the same punishing degree, because that buries the grave decision under the trivial one and grinds the enterprise to a halt. It assigns each decision a **Decision Assurance Level** from its consequence and reversibility, and spends rigor in proportion — the full battery for a one-way door, a light record for a two-way one. Better still, it encodes the common cases as standing rules, so most decisions self-level and human attention is reserved for the ambiguous and the grave. The level is the single dial that tells every other part of the machine how hard to work.

5. Fly the futures — the Decision Flight Simulator

On the gravest decisions, the machine lets you rehearse before you commit. The **Decision Flight Simulator** of Chapter 9 runs the dependency graph forward — *choose A, choose B, or reject, and* *here is the blast radius of each*; sideways — *if the outcome lands here instead of there, which downstream decisions wobble*; and backward — *replay the past to learn from it*. It does not predict the future; it lets you fly the possible ones and walk in ready. Two guards ride on it, always: it **informs, it does not decide**, and it speaks in *reasonable, not absolute* — wearing its assumptions and its uncertainty on its face, because a confident, fabricated future is the most dangerous thing the machine could produce.

6. Assure and seal

When the work is done, the machine produces the thing the whole discipline exists to produce: an **Assurance Record** — the decision, made provably well, to its level — and, where the level earns it, a **Seal of Assurance**. The output is never one thing but two: a grade (*assured to Level B*) and a gap list (*sound here; needs more control there*). And it certifies only what is honest to certify — that the decision was *made* well and can be proven, **never** that it is right. It tracks outcomes, too, but in a separate ledger — the Outcome Ledger of Chapter 8 — and for one purpose only: to calibrate the process over time, never to re-grade a sound decision by how the dice fell.

7. Remember

Finally, the machine keeps what every organization throws away — the memory of Chapter 8, made durable. Each assured decision feeds the corpus, so the institution stops paying for the same mistake twice — the cure for the institutional amnesia of Chapter 1. And the memory compounds: a richer corpus makes the graph sharper and the simulator truer, which makes the assurance more valuable, which earns more decisions worth recording. The asset everyone else discards becomes the one thing a rival cannot copy, because it is your own history.

Above all seven, one rule the machine may never break: **the human stays above the loop.** The tool assures; the person decides and owns the decision. The day the tool decides, it stops being assurance and becomes the oracle this book refused, from the first chapter, to sell.

One decision through the machine

Seven jobs in the abstract can blur together, so let me run a single decision through the machine and show you each one working.

A team proposes adopting a new vendor's artificial-intelligence service to handle a core, customer-facing process. On the team's own reading, it is a routine call — a vendor selection, a Level C, in and out by Friday. Watch what the engine does with it.

It captures the proposal the moment it lands — the recommendation, the options considered, the named owner, the reasons given — as a decision-in-flight, with no one filling out a form. **It grounds** the claims in the proposal, and immediately two go yellow: the headline "ninety-eight percent accuracy" traces only to the vendor's own slide, not to any independent test, and "handles our volume" rests on a demo, not a load test. Neither is a lie; both are simply *ungrounded*, and the machine says so. **It sees the whole**: the dependency graph shows that three already-queued decisions would build directly on this vendor within two quarters, and — the thing no one in the room raised — that a standing data-residency rule, one of the constraints that must never break, conflicts with where this vendor stores its data. **It grades the level**: a customer-facing process that touches personal data, behind a multi-year contract with a data-egress penalty, is not a two-way door. The engine re-grades it from the team's assumed C up to a B, and the easy "in by Friday" path closes. **It flies the futures**: the forward pass shows that adopting the vendor forces two downstream systems to change within six months, and that the residency conflict hardens into a one-way door in about eighteen — the moment the reversibility quietly expires.

Now the part that matters most: at no point did the machine decide. It handed the human auditor a decision that was complete, grounded, graded, and rehearsed — and the auditor, seeing the residency conflict and the unverified accuracy claim, sent it back for an evidence pass and a hosting answer before it could clear. **It assures and seals** the final call: recorded with its rationale, the one engineer's dissent preserved, the evidence attached, sealed at *reasonable assurance, Level B*, with a two-line gap list. **It remembers**: the decision and its expected outcome enter the corpus, where a replay six months on will ask whether the future the team bet on is the one that arrived.

Count what the machine did in the few minutes the team could spare: it grounded forty claims, traced two hundred dependencies, and flew the consequences a year out — work no human committee could finish before the contract needed signing. And count what it caught: the two quiet defects — the ungrounded number and the buried conflict — that without the machine would not surface until eighteen months later, in a budget meeting, as the post-mortem. That is the whole promise, working: not a machine that decided, but a machine that made the deciding sound.

A checklist: what to demand of any assurance tool

When a vendor shows you a "Decision Assurance" or "decision intelligence" platform, hold it to the seven jobs — and walk it through these questions. (The guard rides here too: a tool assures the decision; it does not make it.)

Capture: does recording a decision cost the decider almost nothing — or does it depend on someone filing paperwork later?

Grounding: does every claim, including the tool's *own* outputs, carry checkable evidence — and does the tool abstain when it cannot?

The whole: does it map dependencies and conflicts across decisions, or only score the one in front of you?

Levels: does it grade rigor to consequence, or assure everything the same?

Simulation: if it runs what-ifs, does it show its assumptions and uncertainty — and does it stop short of recommending the answer?

The seal: does it certify the *process* and return a gap list — or does it quietly promise a good *outcome*?

The human: is the person kept above the loop, accountable and in control — or is the tool the decider in disguise?

A tool that fails questions two, six, or seven is not assurance. It is a decider wearing a clipboard. *(This joins the Field Kit as the tool-evaluation checklist.)*

"But isn't this just a decision support system?"

If you spent time in enterprise technology, an objection is forming, and it deserves a straight answer: *we built this before — the decision support system, the executive information system, and they ran what-if simulations decades ago.* They did, and I am the last person to wave them away; I wrote my graduate thesis on the executive kind. But they answer a different question at a different altitude. A decision support system lives *inside* the loop: it feeds the decider a better answer at the moment of choice, and when the choice is made, the analysis evaporates. Decision Assurance lives *above* the loop: it makes the answer one you can *prove* — complete, grounded, recorded, defensible long after everyone moved on. The support system asks *what should we do?* Assurance asks *can we show we decided this well, and will the proof outlive us?*

The difference is not academic, and one famous number shows why. In 2010, two respected economists published a study with a striking finding: when a country's public debt climbs past ninety percent of its annual output, growth does not merely slow — it turns sharply negative, as if the economy steps off a cliff. The paper landed in the middle of the worst downturn in eighty years, and finance ministers across continents reached for it. It became one of the most-cited justifications for austerity — for cutting budgets, services, and jobs — on the argument that crossing the ninety-percent line invited catastrophe. The what-if spoke, and governments acted.

Three years later, a graduate student trying to reproduce the result asked for the authors' original spreadsheet. They sent it. And there, in the formula that computed the headline average, sat an error a first-year analyst would recognize: the calculation reached across only fifteen of the twenty countries in the data, silently dropping the other five from the average. Correct the formula, restore the missing rows, and the cliff softened into a gentle slope. The dramatic threshold that helped justify a decade of hard policy was, in meaningful part, a spreadsheet that averaged the wrong rows.

The world was awash in decision support and still walked through a one-way door on an ungrounded number. Support handed everyone the what-if. Nobody handed them assurance — no independent check that the model behind the number was sound, no record that anyone looked. So yes, the engine runs what-ifs — that part it shares with the old tools, and it is the least of what it does. The hard part is everything wrapped around the what-if: grounding the inputs, assuring the model itself, recording the run, an independent check, and a later look at whether the rehearsed future actually arrived. We do not claim a new kind of what-if. We claim the first one you can trust.

An honest accounting — the bar, not the brochure

Everything above is the *ideal* tool, and an ideal is a standard you hold yourself to, including your own attempts. So let me be plain about the distance between the bar and today's technology — because a book about assurance that oversold its own machine would fail its first test.

The dependency graph is only as honest as its edges, and a model will invent an edge as fluently as it invents a fact; a real system must flag where it is guessing and present a map with its holes drawn in, or it manufactures a false sense of completeness that is worse than none. The simulator is a model, and every model is wrong and some are useful; its futures are reasonable, never certain. Grounding sharply reduces fabrication — in our own pre-registered work, requiring each claim to cite its evidence roughly halved it — but it does not reach zero, which is exactly why the human stays above the loop and the seal says *reasonable*, not *absolute*. And the board of agents does not out-decide the expert in the chair; Chapter 3 settled that, and nothing here un-settles it.

Job by job, here is the candid scorecard — what stands up today, and where the work still is.

The job	**What stands up today**	**The honest gap**
Capture	a model can pull a decision out of a meeting or a thread	getting people to let it, cleanly, without friction
Ground	retrieval can attach evidence; "cite or abstain" is buildable	the model still fabricates citations; grounding halves, not erases
See the whole	a model can extract decisions and propose the links	the graph invents edges; a false "complete" is the worst output
Grade the level	a rules engine plus a few questions levels most	the ambiguous cases need judgment, and people

	decisions well	game the level down
Simulate	forward propagation over a known graph is tractable	reasonable, never absolute; garbage in; the branches explode
Assure & seal	a structured record, a grade, and a gap list generate cleanly	the seal means nothing until the profession earns trust — years
Remember	storage is the easy part; the corpus compounds	the politics and privacy of a graph that exposes who decided what

Read the right-hand column and you notice the gaps cluster in the same place: they are precisely the spots where human judgment, independence, and trust stay essential. That is not a coincidence to engineer away. It is the shape of the discipline. Anyone who tells you the whole of it is solved today is selling you the oracle.

So here is the claim, sized to what the evidence supports. The ideal tool does not make better decisions than your best people. It makes the decisions your people already make **complete, grounded, graded, recorded, rehearsable, and provable** — at a scale, and a cost, that for the first time in the history of the problem is finally within reach. That is not a smaller promise than the oracle. It is the only one worth trusting.

So what?

For you, the seven jobs are a lens you can hold over anything that calls itself a decision tool — your current stack, a vendor's demo, a thing you might build. Most will light up two or three of the seven and go dark on the rest, and the dark ones tell you precisely where your assurance gap lives. The machine is not magic; it is the patient automation of the discipline, pointed at the one output your enterprise never instrumented.

What now?

Take the tool-evaluation checklist to the next "AI for decisions" pitch that lands on your desk, and score it, question by question, in front of the people selling it. Watch what happens at questions two, six, and seven — grounding, the seal, and the human. The honest vendors lean in. The rest change the subject. Either way, you will know in ten minutes whether you are looking at assurance or at a decider in a clipboard.

Where we go next

We now hold the discipline, the levels, the profession, and the shape of the machine that serves them. What remains is the most practical question of all: how an ordinary organization stands this up on an ordinary Monday, without boiling the ocean. That is the next chapter.

Chapter 10 in review

Key takeaways

- The tool is **not** the discipline — it gives the human auditor *reach* (throughput and synthesis), never replacing the judgment that stays above the loop.
- A Decision Assurance system does **seven things**: capture the decision, ground every claim, see the whole (the dependency graph), grade the level, fly the futures (the simulator), assure and seal, and remember.
- **Capture must be nearly free**, or the discipline dies at step one; and the machine's *own* outputs must be grounded, because the assurer can hallucinate as fluently as the decider.
- The **dependency graph** is how the machine watches the whole that no human committee can hold — completeness and governance at scale.

The engine **grades rigor to consequence** (the level as the master dial) and, on the gravest calls, lets you **rehearse the futures** — informing, never deciding; reasonable, never absolute.

It is **not a decision support system**: support sits in the loop optimizing the answer; assurance sits above it, proving the process — the same what-if, wrapped in grounding, a record, an independent check, and a seal.

Honest accounting: the ideal tool does not out-decide your people (Chapter 3 stands); it makes their decisions complete, grounded, graded, recorded, rehearsable, and provable — affordably, for the first time.

Traced through one decision, the seven jobs show the payoff: the machine grounds the claims, maps the dependencies, and flies the futures faster than the team can in the time available — catching the quiet defects (the ungrounded number, the buried conflict) that would otherwise surface only as the post-mortem — while the human still decides.

Key terms

The seven jobs — capture · ground · see the whole · grade the level · simulate · assure-and-seal · remember; the functional spine of a Decision Assurance system.

Tool-evaluation checklist — the seven jobs turned into buyer's questions; a tool that fails grounding, the seal, or the human is a decider in disguise.

Grounded outputs (of the tool) — the rule that the assurance machine cites checkable evidence for its *own* assertions and abstains otherwise — the in-tool answer to "who assures the assurer?"

The bar, not the brochure — the ideal tool as a standard held over every system, including the author's own; paired with an honest statement of how far today's technology falls short.

The honest scorecard — a job-by-job statement of what assurance technology does well today and where it still falls short; the gaps cluster exactly where human judgment, independence, and trust stay essential.

Check your understanding

Why is "capture the decision as it happens" the first job, and what fails if it is missing?

Give the rule the machine must hold for grounding — and why it applies to the tool's own outputs.

What does the dependency graph do that a human committee cannot?

Name the two guards that ride on the Decision Flight Simulator.

In one line, how does this differ from a decision support system?

What does the "honest accounting" refuse to claim, and what does it claim instead?

In the worked trace, what two defects did the machine catch that the team's own reading missed?

Answer key: (1) A decision is perishable; without near-free capture, nothing downstream exists to assure, and the discipline dies at step one. (2) Every claim cites checkable evidence or is withheld; it applies to the tool's own outputs because the assuring model can hallucinate as fluently as the deciding one. (3) Hold thousands of interdependent decisions at once and propagate a change — antecedents, conflicts, downstream impact — without tiring or forgetting. (4) It informs but does not decide; it speaks in reasonable, not absolute (showing assumptions and uncertainty). (5) A support system sits in the loop optimizing the answer;

assurance sits above it, proving the process. (6) It refuses to claim the machine out-decides your best people; it claims it makes their decisions complete, grounded, graded, recorded, rehearsable, and provable — affordably. (7) The ungrounded accuracy claim (traced only to the vendor's slide, not an independent test) and the buried data-residency conflict (a constraint that must never break).

Questions for reflection and discussion

Run the seven jobs over the decision tooling you use today. Which jobs light up, and which are dark — and what does the pattern of dark ones tell you?

Where would near-free *capture* change your organization most — and what makes recording a decision feel like work today?

For facilitators: give the room a real "AI for decisions" vendor page and ask them to score it against the seven questions; the disagreements reveal what each person quietly believes "assurance" means.

Apply it

Pick one decision system in your organization — a governance workflow, an Architecture Review Board tool, an "AI copilot," a spreadsheet that drives real calls — and grade it against the seven-question checklist, marking each *yes*, *partial*, or *no*. Then write one sentence per *no* describing the assurance gap it exposes. The result is a one-page honest scorecard of your current state, and a build-or-buy list ordered by what is missing. *Trainers:* ask two groups to score the same system independently and compare; where they disagree is where your organization never pinned down what it means to assure a decision.

Notes and further reading

This chapter synthesizes the framework of Chapters 1–9: the five properties (Ch 4), the lineage and independence (Ch 5), the Decision Assurance Levels (Ch 6), the auditor, the Decision Audit, and the Seal of Assurance (Ch 7), the two ledgers (Ch 8), and the dependency graph and simulator (Ch 9).

The decision-support distinction draws on Gorry & Scott Morton (1971) and Rockart's executive information systems (1979, 1982); the austerity-model case is Reinhart & Rogoff (2010), the spreadsheet error identified by Herndon, Ash & Pollin (2013). Fuller treatments live in the Story Banks and the standalone "Isn't this just a decision support system?" beat.

The "grounding roughly halved fabrication" figure and the "a board does not out-decide its best agent" result are from the book's own pre-registered research (OSF, DOI 10.17605/OSF.IO/ZWM3S); the *reasonable, not absolute* humility is from Chapter 5.

CHAPTER 11
Monday Morning

> *Type 2 decisions can and should be made quickly. — Jeff Bezos, 2015 Letter to Shareholders*

You close the book convinced. Then it is Monday morning, and conviction meets the calendar.

This is the moment most good ideas die — not in disagreement, but in the gap between *I believe this* and *I know what to do at 9 a.m.* You run a real organization. You own a packed week, a skeptical change board, and exactly as much political capital as you spent last quarter and no more. So the honest question is not whether Decision Assurance is worth doing. You settled that somewhere back in Part II. The question is where a busy person *starts* — with the resources you actually command, not the ones a transformation program would require.

Here is the whole answer, and it will sound almost too small: you do not roll out Decision Assurance. You start it. With one decision. This week.

Let me show you why the small start is the only one that works, and exactly which steps to take before lunch.

The death most governance dies

First, the trap — because you will feel its pull by Tuesday.

The instinct of every capable organization, handed a good framework, is to do it *properly*: charter a working group, draft a forty-page policy, select a tool, build a steering committee, and schedule the eighteen-month rollout. It feels responsible. It feels thorough. And it is how most governance dies — not with a rejection, but with a binder. Too much, too formal, too soon. By the time the policy is ratified, the people it governs route around it, the tool sits unused, and "Decision Assurance" enters the company vocabulary as a synonym for red tape. You will recognize the corpse, because it is the same one the Change Approval Board became: a ritual everyone performs and no one believes.

This is also where the skeptic in the third row raises a hand: *isn't this just a glorified spreadsheet — more process for its own sake?* The skeptic is not wrong to worry. A heavy, ceremonial rollout *would* be exactly that. The answer is not to argue the skeptic down. It is to refuse to build the thing they fear.

The minimum viable practice

So borrow a move from the people who build products under uncertainty.

When Frank Robinson named the *minimum viable product* in 2001 — the idea Eric Ries later carried into *The Lean Startup* in 2011 — the point was not to ship something flimsy. It was to stop betting everything on a grand plan you cannot yet validate, and instead build the smallest real thing that teaches you whether you are right. Decision Assurance deserves the same humility. Forget the program. Build the **minimum viable practice**: assure *one decision*, completely, this week.

One decision, done fully — captured, grounded, graded, sealed, remembered — is worth more than a policy that covers all of them on paper and none of them in practice. It is real. It produces an artifact you can hold up: *here is what an assured decision looks like in our organization, with our names on it.* It convinces the way no slide deck convinces, because it already happened. And it costs you a single afternoon, not a single budget cycle.

The sequence, before lunch

Here is the Monday-morning sequence. You already own every concept in it — this chapter only puts them in order and sets them in motion.

Pick the decision. Not just any decision — the right one. Reach for a call with a large blast radius and a door that does not reopen. Jeff Bezos drew the line cleanly in 2015: some decisions are reversible two-way doors you walk back through if you dislike the other side, and some are one-way doors you cannot. Start with a one-way door — high consequence, low reversibility, exactly the kind the Decision Assurance Levels reserve their rigor for. Assuring a trivial decision proves nothing. Assuring the migration, the reorganization, the irreversible bet proves everything.

Capture it. Write down the question on the table, the options considered, the evidence in hand, and the name of the person who owns the call. Plainly, in minutes. Capture is the foundation, and it must be nearly free, or it will not happen — more on that in a moment.

Ground it. Tie every load-bearing claim to a source. Not "the market is moving this way" but *which* signal, from *where*. Grounding is what separates a decision from an opinion wearing a decision's clothes.

Grade the level. Run the quick triage from the Field Kit: how severe, how reversible? The answer sets how much assurance this decision earns — and, just as importantly, tells you when to *stop*. Most decisions deserve less rigor than you fear; the grave ones deserve all of it.

Record and seal it. Assemble the above into an Assurance Record and seal it — the immutable snapshot from the Memory chapter, the thing that answers the audit email eight months early. It is now a fact in your organization, not a memory in your head.

Log the outcome — later, and separately. When the result lands, record it in the Outcome Ledger. Never reach back to re-grade the sealed decision by its outcome. Outcomes calibrate the next decision; they do not rewrite the last one.

Do the next one. Then the one after. This is the whole growth model: not a launch, but a practice that compounds, one assured decision at a time.

Make capture nearly free

Step two carries the whole sequence, so it deserves its own warning: if recording a decision costs more than making it, no one will record anything, and your beautiful practice will be empty by March.

The failure mode is the form — the fourteen-field template someone fills out, resentfully, a week after the decision, from memory, to satisfy a process. That is not capture. That is archaeology. Real capture happens *at the moment of decision*, in the flow of the meeting or the message thread where the call actually gets made, with the lightest possible touch — a few structured lines, not a document. The discipline is to drive the cost of capturing a decision toward zero, because every ounce of friction you leave in is a decision someone will quietly decline to record. Cheap capture is not a nicety. It is the difference between a living ledger and an empty one.

From practice to policy

Once you assure one decision, then ten, then a hundred, a question arrives on its own: who makes sure the rules are applied the same way every time, when you are not in the room? The answer is to stop enforcing the rules by memory and start enforcing them by machine.

The software world solved a version of this and named it well. *Infrastructure as code* taught engineers to express their systems as instructions a machine runs, identically, every time, instead of as a runbook a tired human follows at 2 a.m. *Policy as code* extended the idea to rules — encoding a policy so the system enforces it automatically and consistently, with tools like the Cloud Native Computing Foundation's Open Policy Agent and HashiCorp's Sentinel now standard in the field. Decision Assurance reaches for the same principle at the level of *governance* — the broader frame I call **governance as code**: the assurance levels, the evidence each one requires, the sign-offs each one triggers, all expressed as policy the system runs, rather than guidance a busy reviewer is trusted to remember.

This is what the architects among us mean by a *constitution*: not a document on a shelf but the encoded rules your assurance practice actually follows — visible, versioned, and applied identically to every decision, including the ones made when the most conscientious person on the team is on vacation. You do not need it on Monday. You will want it by the time the practice outgrows your personal attention, which arrives sooner than you expect.

Immutable does not mean uncorrectable

One discipline to set down now, because it prevents a quiet corruption later.

A sealed Assurance Record is immutable — that is the point of it; a record you can quietly rewrite is a record that proves nothing. But real records sometimes carry a real error: a figure transposed, a source misattributed, a date wrong. The discipline is not to edit the sealed record in silence, and it is not to pretend the error never occurred. It is to **append a correction** — dated, signed, and itself recorded — that sits alongside the original and sets it right. Immutable means *uncorrupted,* not *uncorrectable*. The correction is part of the record, the way an erratum is part of the literature: the trail stays honest precisely because nothing is erased. Build this habit on day one, and you never face the temptation to reach into history and tidy it.

Who owns it

A practice needs an owner, and on Monday morning the owner is you.

Eventually, Decision Assurance belongs to a role — the decision auditor we met earlier, the steward who runs the practice, maintains the levels, and holds the seal. But you cannot hire that role before you prove the practice is worth one, and you will wait forever if you make the org chart a precondition. So you start as the owner yourself: the chair, the architect, the person who already feels the gap most sharply. You run the minimum viable practice by hand, you accumulate the assured decisions, and you let the evidence make the case for the role. The sequence is practice first, role second — never the reverse. Standing it up does not wait on permission. It waits on you.

Do not rebuild the board you hated

A warning, because zeal is the enemy here.

The moment Decision Assurance starts working, you will feel the urge to assure *everything* — to put every decision through the full rigor, because rigor feels like virtue. Resist it. An organization that applies its heaviest process to its lightest decisions reinvents the Change Approval Board, the very ceremony you set out to escape. Bezos named the pathology precisely: as organizations grow, they tend to force the slow, one-way-door process onto fast, two-way-door decisions, and the result is sluggishness, timidity, and a slow strangling of initiative. The Decision Assurance Levels exist to prevent exactly this. Most decisions stay fast and unceremonious. Only the one-way doors earn the full treatment. Proportion is not a compromise of the discipline. It *is* the discipline.

Ninety days, in real time

Let me make the compounding concrete, because "a practice this quarter" is easy to say and hard to picture. Here is the quarter.

Week one. A real decision lands on your desk: consolidate two overlapping data platforms onto one, retiring the other. A one-way door — once the data moves and the old system goes dark, there is no walking it back. Exactly the kind the levels reserve their rigor for. You run the sequence by hand. Twenty minutes to capture the question, the three options, and the evidence. Ten more to ground the load-bearing claims — the migration cost estimate traced to the team that made it, the "the old platform is end-of-life" assertion checked against the vendor's actual notice. You grade it Level B, seal the record, and name yourself the owner. Before lunch, your organization owns its first assured decision. It cost you half a morning and no budget.

Week four. Six more followed — not because a policy required it, but because the first one felt good and the second one was faster. A pattern is forming. The capture that took thirty minutes in week one now takes eight, because you stopped writing documents and started writing a few structured lines in the moment. People notice the records. A colleague, heading into a hard call of her own, asks: *can you send me the format you used?* That sentence is the practice spreading on its own — no mandate, no rollout, just a useful thing being copied because it is useful.

The skeptic. Somewhere in week five, the change-board skeptic — the one who called it a glorified spreadsheet — watches a sealed record settle an argument in four minutes that would once take a week of "who decided this?" And the objection quietly changes shape. Not *why are we doing this,* but *can we use this for the platform decision too?* You did not win the argument. The artifact won it for you.

Month three. You hold something now that did not exist ninety days ago: a small stack of assured decisions, sealed, with their expected outcomes pre-registered. And the first outcomes are landing. One bet paid off as predicted; one came in worse, and the Outcome Ledger catches the delta. You run your first calibration — fifteen minutes, comparing what you expected against what happened across a dozen decisions — and a pattern surfaces that no single decision could show: your team's cost estimates run optimistic, consistently, by a similar margin. That single insight, impossible without the record, sharpens every estimate the team makes next quarter. The practice starts to teach you about yourselves.

None of this required a program. No working group, no forty-page policy, no eighteen-month timeline. One person, one decision, one Monday — repeated until it became the way things are done. The quarter did not transform the culture; cultures do not turn in a quarter. But the engine is running, the evidence is accumulating, and the case for the role, the tool, and the policy now makes itself, because the practice already works. That is how the small start beats the grand plan: not by being more ambitious, but by being real.

The long view

Let me be plain about the timeline, because I would rather you start with the right expectation than abandon the practice when it does not transform your culture by Friday.

This takes years to become who you are. One assured decision this week. A working practice this quarter. A named role this year. A culture — the kind where no consequential decision goes unassured and no one thinks twice about it — over several. That pace is not a failure of ambition. It is how durable things are built. I run marathons the same way I run governance work: not by sprinting, but by building the engine and running until the destination arrives. The training that lasts is the kind that trains the tired leg, not the fresh one — the patient mileage, the unglamorous Tuesday. Enterprise change is the same. The organizations that institutionalize Decision Assurance will not be the ones that launched it the loudest. They will be the ones that started smallest and never stopped.

So start small. Start Monday. Assure one decision before lunch, and let it compound.

The honest caveats

This practice fails in predictable ways, and naming them is how you avoid them.

It can become theater — adopted for the look of rigor, with a ledger no one reads and a level no one enforces. The cure is to start with a decision that truly matters, not a safe pilot no one cares about; real stakes keep a practice honest.

It can become bureaucracy — if you forget proportion and assure everything, you rebuild the board you escaped. The levels are the antidote; use them to say *no* to rigor as often as *yes*.

It can die of friction — if capture costs more than deciding, the ledger empties. Guard the cost of capture as the single most important number in the practice.

And it can trigger the politics of autonomy — because governance, to the person being governed, can feel like surveillance. Lead with the truth that disarms it: an Assurance Record protects the good decider. It is the thing that defends a sound call when the outcome turns and the second-guessers arrive. Sold as protection rather than policing, assurance is something people ask for, not something they evade.

So what?

You do not launch Decision Assurance; you start it — with one consequential decision, fully assured, this week — and let it compound into a practice, a role, and eventually a culture, with the Decision Assurance Levels keeping it proportionate the whole way.

What now?

Before your next meaningful decision, run the seven-step sequence on it by hand — pick, capture, ground, grade, record and seal, log the outcome later, then do the next one. One decision is the entire beginning.

Where we go next

You now know the discipline, the levels, the auditor, the memory, the graph and simulator, and how to stand it all up. One question remains, and it is fair to ask the author directly: *you built one of these — how does it actually score?* The next chapter answers, plainly, against the seven jobs.

The chapter, in review

Key takeaways

Decision Assurance dies the way most governance dies — in a **big-bang rollout** that becomes a binder no one reads; you start it instead of rolling it out.

Build the **minimum viable practice**: assure *one* consequential decision, completely, this week — one real assured decision outweighs a policy covering all of them on paper.

The Monday sequence is seven steps: **pick** (a one-way door), **capture**, **ground**, **grade the level**, **record and seal**, **log the outcome later** (never re-grade), and **do the next one**.

Make capture nearly free — capture at the moment of decision, in minutes; friction is what empties a ledger.

As the practice grows, move from rules-by-memory to **governance as code** — the levels, evidence, and sign-offs encoded as policy the system runs (the practical meaning of a "constitution").

Immutable does not mean uncorrectable: never edit a sealed record in silence; append a dated, signed correction that sits beside the original.

Ownership starts with you (the chair, the architect) and evolves into the decision-auditor role — practice first, role second, never waiting on the org chart.

Keep it **proportionate** — assure the one-way doors, let the two-way doors stay fast — or you rebuild the Change Approval Board you escaped; the long view expects years, not weeks.

Key terms

Minimum viable practice — the smallest real implementation of Decision Assurance: one consequential decision assured completely, used to prove and grow the discipline rather than launch it all at once.

Governance as code — expressing the assurance rules (levels, required evidence, sign-offs) as policy the system enforces automatically and consistently, rather than guidance a reviewer is trusted to remember; the broader cousin of "policy as code."

Append a correction — the discipline that keeps an immutable record honest: errors are set right by a dated, signed addition recorded alongside the original, never by silent editing.

One-way / two-way door — Bezos's distinction between irreversible decisions (decide slowly, assure fully) and reversible ones (decide fast, assure lightly); the practical trigger for how much rigor a decision earns.

Check your understanding

Why does a big-bang governance rollout tend to fail, and what does "start it, don't roll it out" mean in practice?

Walk through the seven steps of the Monday sequence in order.

Why is the *cost of capture* described as the single most important number in the practice?

Explain "immutable does not mean uncorrectable." How is a real error in a sealed record handled?

How do the Decision Assurance Levels keep the practice from becoming the bureaucracy the skeptic fears?

Answer key. (1) It is too much, too formal, too soon — by the time the policy ratifies, people route around it and it becomes red tape; "start it" means proving the discipline on one real decision rather than launching a program. (2) Pick a one-way-door decision; capture it; ground every claim; grade its level; record and seal it; log the outcome later without re-grading; do the next one. (3) Because if capturing a decision costs more than making it, no one will, and the ledger empties — friction is the practice's chief enemy. (4) Immutable means the sealed record cannot be quietly rewritten, but a genuine error is corrected by appending a dated, signed correction beside the original, so nothing is erased and the trail stays honest. (5) By matching rigor to consequence — only one-way doors earn the full treatment, while reversible decisions stay fast — so assurance never becomes heavy process applied to everything.

Questions for reflection and discussion

What is the single most consequential, least reversible decision on your desk right now — and what would it take to assure just that one this week?

Where in your organization does capturing a decision currently cost more than making it, and what friction could you remove?

For facilitators: Ask the group to name a past governance initiative that became "a binder no one reads." What made it heavy, and how would a minimum-viable-practice start change its fate?

Who, realistically, would own the assurance practice in your organization on Monday — and what evidence would justify making it a formal role?

Apply it

Run the minimum viable practice for real. Pick one truly consequential decision currently in front of you, and walk it through all seven steps this week — capturing in minutes, grounding every claim, grading its level, and sealing the record. Keep a note of where the practice felt heavy and where it felt natural. *Trainers:* ask each participant to bring their sealed record to the next session and present it in two minutes; the contrast between the records that took ten minutes and the ones that took two is the clearest lesson available in what "make capture nearly free" actually means.

Notes and further reading

The minimum viable product. Term coined by Frank Robinson (2001); popularized by Eric Ries, *The Lean Startup* (Crown Business, 2011), within the build-measure-learn loop. ("Minimum viable practice" adapts the spirit —

the smallest real thing that proves the idea — to a governance discipline.)

Policy as code. An extension of "infrastructure as code"; representative tooling includes the Cloud Native Computing Foundation's Open Policy Agent (CNCF-graduated) and HashiCorp Sentinel. "Governance as code" is used here as the broader framing at the level of the assurance discipline.

Reversible and irreversible decisions. Jeff Bezos, Amazon.com 2015 Letter to Shareholders: "Type 1" one-way doors (irreversible — decide methodically) versus "Type 2" two-way doors (reversible — decide quickly); the warning that large organizations over-apply the heavy process to light decisions.

CHAPTER 12

I Built One of These

> *The first principle is that you must not fool yourself — and you are the easiest person to fool. — Richard Feynman, "Cargo Cult Science," 1974*

For most of this book, I kept the tool out of sight. That was deliberate.

A book about a discipline should not double as a brochure for a product, and the fastest way to lose your trust would be to spend ten chapters defining a standard and then reveal that the whole exercise was a runway for the thing I happen to sell. So I left it offstage. The seven jobs, the levels, the auditor, the two ledgers, the flight simulator — all of it stands on its own, whether or not anyone ever builds the tool I built, and you can apply every word of it to a spreadsheet, a competitor, or a practice you assemble by hand.

But fairness cuts both ways. In the last chapter I asked you to take the seven jobs and grade every tool that courts you — including, I said, the author's own. It would be a small cowardice to send you off to grade other people's tools and never submit mine to the same ruler. So here it is, named at last and measured in public, gaps and all. The tool is called **Arclave.** Let me show you exactly where it stands today — what works, what half-works, and what is still a promise rather than a fact.

What it is, in one breath

Arclave began as a question I could not stop asking: what would it look like to give an Architecture Review Board a working twin — a system that does the assurance jobs around a real decision, so the humans can do the deciding?

That is all it is, and all it claims to be. It is an attempt to make this book's discipline concrete: to capture a decision, ground it in evidence, hold it to a level, record and seal it, and remember it — under a written constitution, with the reasoning visible the whole way through. It is not an oracle, and the most important thing I can tell you in this chapter is that I can prove it is not one, because I tried to find out and the answer was no. We will come to that. First, the scorecard.

The honest scorecard

Here is Arclave against the seven jobs from the ideal-tool chapter, today, with no rounding up. Three marks: **Built** (it works at the level of the current Decision Console), **Partial** (some of it works, some is scaffolding), and **Not yet** (it is design and intent, not a working capability).

#	The job	Where Arclave stands	Mark
1	Capture the decision	The Decision Console captures the question, the board's findings, and the call, in the flow	Built
2	Ground every claim in evidence	Findings are evidence-cited; a standing "constitution lane"	Built

		checks each against the rules	
3	See the whole (the dependency graph)	The model exists on paper; the working graph of decision dependencies is not built	Not yet
4	Grade the level (Decision Assurance Levels)	The framework is defined and the console renders a verdict with conditions; full leveling-as-policy is partial	Partial
5	Simulate (the flight simulator)	Designed in detail; the what-if engine is not built	Not yet
6	Assure and seal	The console issues a verdict, rationale, and conditions, and produces a sealed, branded Assurance Record	Built
7	Remember	A decision ledger persists each record; the second ledger — the outcome-calibration loop — is not yet wired	Partial

Read that table closely and a pattern jumps out, and it is not a flattering one: **the two jobs marked *Not yet* are the two that sound the most impressive.** The dependency graph and the flight simulator — the marquee, the chapter I told you was the top of the value — are precisely the parts Arclave cannot yet do. What works today is the quieter half: capture, grounding, sealing, and the beginnings of leveling and memory. The assurance plumbing is real. The crystal ball is still a drawing.

I could dress that up. I will not, because a Decision Assurance tool that fudged its own scorecard would fail the first test of its own discipline.

The hardest line in the book

Now the admission that matters more than the table, because it is the one most founders would bury.

When I started Arclave, I believed — and I designed it to demonstrate — that a board of specialized artificial-intelligence agents, deliberating together, would decide *better* than its best single member. A team of minds, each catching what the others missed, converging on a wiser answer than any one of them could reach alone. It is an attractive idea. It is the idea most of the industry is quietly selling. And it is the idea this book is named against, because I went and tested it on myself.

I wrote the hypothesis down *before* running the experiment — pre-registered it publicly on the Open Science Framework, so I could not later move the goalposts — and then I ran it. Across seven different experimental regimes, the result came back the same way each time: **the board did not beat its best single calibrated agent.** Not when I de-correlated the agents across model families. Not when I added a devil's advocate. The super-additive intelligence I expected was not there, and I understand now why — the gains wash out in how the aggregate sets its accept-and-reject thresholds. I posted the disconfirming result alongside the pre-registration, where anyone can read it.

That is the least comfortable paragraph I will write in this book, and it is also the most important, because it is the empirical floor under everything I asked you to believe. Arclave does not out-decide your experts. No assembly of agents I could build out-decided the single best mind in the room. What the system *does* — reliably, measurably — is assure: it makes the deciding complete, grounded, graded, sealed, and remembered, and it keeps the human in the chair. I did not learn the difference between assurance and an oracle from a whiteboard. I learned it from a hypothesis of my own that did not survive contact with the evidence. Feynman's line is the truest sentence I know about this work: the easiest person to fool is yourself, so the first discipline is refusing to.

The scorecard is the point, not the product

Step back from Arclave for a moment, because the reason this chapter exists is not to sell you the tool. It is to hand you the ruler.

Look again at what we just did. We took a tool — mine — and held it against seven named jobs, marked each one Built, Partial, or Not yet, and refused to launder the gaps. That is the whole method, and it works on any tool, not just this one. The vendor who arrives next quarter with an "AI decision platform" can be set against the same seven rows in ten minutes. Where do they capture? Where do they ground? Can they show you the dependency graph, or is it a slide? Does the thing seal a record you could hand an auditor, or does it just print a confident answer? The seven jobs are the ruler. Arclave is simply the first thing I measured with it, and I measured it in public so you would believe the ruler is real.

This is the deeper reason I built the scorecard into the book rather than into a sales deck. A sales deck grades to impress. This grades to inform — which is the entire difference between a brochure and an Assurance Record. The table above *is* an Assurance Record: a claim about Arclave, graded by level, with the evidence and the gaps on the page. I assured my claims about the product the same way I am asking you to assure your decisions. Anything less would be hypocrisy with a logo.

What is proven, and what is only promised

Two more honesties, so you know exactly how far the evidence reaches.

The corpus is prospective. Arclave's assurance is demonstrated on a controlled board working real decisions in a deliberate, instrumented setting — not yet on a large body of live enterprise decisions across many organizations. The mechanism works where I can watch it work. Whether it holds at the scale and mess of a Fortune 100 portfolio is a claim I am still earning, not one I will pretend is settled.

And the most interesting question is still open. There is a second study — already designed, already pre-registered in draft — that asks the one thing that would matter most: not whether the board decides *better* than a lone expert, but whether it *catches things* a lone expert misses. Novel defects. The quiet flaw the single reviewer, however brilliant, did not happen to look for. That is the test I most want to run, and the status is plain: it is designed, and **not yet run.** Until I do, the "catches what you would miss" claim stays where it belongs — in the column marked *promising,* not the column marked *proven.*

I would rather tell you that plainly now than let you discover the gap on your own and wonder what else I rounded up.

The invitation

So here is what I actually want from this chapter, and it is not your purchase order.

Take the seven jobs. Take the three marks — Built, Partial, Not yet. And run them on every tool that asks for your trust, starting with the most confident one in the room. Watch what happens when you ask a vendor to grade their own product in front of you and to name the two jobs it does worst. The ones who can do that — who show you the gaps before you find them — are the ones worth your time. The ones who claim all seven, fully, today, with no Partial and no Not yet, are selling you the oracle this whole book spent its length refusing, and the most useful thing you can do is thank them and leave.

Grade Arclave that way too. I built it, I believe in it, and I just showed you the two biggest things it cannot yet do and the hypothesis of mine that the evidence killed. That is the standard. Hold every tool to it, including this one, especially this one.

The honest caveats

This chapter carries risks I should name, in keeping with the rest.

A self-assessment is not an independent one. I graded my own tool, and a tool's maker is exactly the person most likely to fool himself about it — Feynman's warning applies to me first. So treat this scorecard as a worked example of the method, not as assurance in the full sense; the real thing requires a second grader who does not share my hopes. The discipline's own logic demands an independent eye, and I am not it.

A scorecard is a snapshot, and snapshots age. Two of those *Not yet* marks will, I intend, become *Partial* and then *Built*; if you are reading this a year on, re-grade it yourself rather than trusting the marks I printed today. A claim about a moving product is only as current as its date.

And the reveal itself carries a pull toward the pitch. Naming Arclave after ten chapters of discipline risks reading as the brochure I promised this was not. My only guard against that is the one I chose on purpose: I led with the gaps and the disconfirmed hypothesis, not the features. If this chapter still reads like a sales page to you, trust the gaps over the prose — they are the part a sales page would never include.

Measure the tool. Measure mine first. And keep the deciding where it belongs.

So what?

Held against its own seven jobs, Arclave today is **Built** on capture, grounding, and sealing; **Partial** on leveling and memory; and **Not yet** on the dependency graph and the flight simulator — and its own pre-registered experiment disconfirmed the hope that a board of agents out-decides its best member, which is exactly why it offers assurance rather than an oracle.

What now?

Take the seven-job scorecard and grade one tool you already use or are being sold — marking each job Built, Partial, or Not yet, and naming the two it does worst. The exercise tells you more in ten minutes than a vendor demo tells you in an hour.

Where we go next

That is the discipline, the framework, the practice, and an honest look at one attempt to build it. The last chapter lifts the lens all the way up — from a single decision, and a single tool, to the decisions of a civilization.

The chapter, in review

Key takeaways

The tool stayed offstage for ten chapters on purpose; naming it now, and grading it by the same ruler, is a matter of fairness, not promotion.

- **Arclave** is an attempt to make the discipline concrete — an Architecture Review Board's working twin that does the assurance jobs so the humans can decide.
- The **honest scorecard**: Built on capture, grounding, and sealing; Partial on leveling and memory; **Not yet** on the dependency graph and the flight simulator — the two most impressive-sounding jobs are the two least built.
- The author's own pre-registered experiment **disconfirmed** the hope that a board of agents out-decides its best single member, across seven regimes — the empirical floor under the whole book's "assurance, not oracle" claim.
- The point is the **ruler, not the product**: the seven jobs grade any tool, and Arclave is simply the first one measured, measured in public.
- The scorecard **is itself an Assurance Record** — a claim graded by level with evidence and gaps shown — because a Decision Assurance tool that hid its gaps would fail its own test.
- What is proven is bounded: the corpus is **prospective**, and the most important study (does the board *catch* what a lone expert misses?) is designed but **not yet run** — promising, not proven.
- The invitation: grade every tool this way, especially the most confident one; the vendor who shows you the gaps first is the one to trust.

Key terms

Arclave — the author's Decision Assurance tool: an Architecture Review Board digital twin built to perform the assurance jobs around a real decision while the human keeps the decision.

The honest scorecard (applied) — a tool graded against the seven jobs with three marks (Built / Partial / Not yet), gaps shown; here, the author's own tool, graded in public.

Pre-registration — writing down a hypothesis and method publicly *before* running the experiment, so results cannot be quietly reinterpreted after the fact; the safeguard behind the disconfirmed-board finding.

Prospective corpus — assurance demonstrated in a controlled, instrumented setting rather than across a large body of live decisions; a claim still being earned, not yet settled.

Check your understanding

Why did the author keep Arclave offstage for most of the book, and why name it now?

Which two of the seven jobs does Arclave not yet do — and why is it notable that those are the two?

What did the author's pre-registered experiment test, and what did it find?

Explain why the scorecard is described as "itself an Assurance Record."

What is the difference, in this chapter, between what Arclave *proved* and what it only *promised*?

Answer key. (1) To keep the book a discipline rather than a brochure; it is named now because fairness requires submitting the author's own tool to the same seven-job ruler the reader was asked to use. (2) The dependency graph (see the whole) and the flight simulator (simulate); notable because they are the most impressive-sounding capabilities yet the least built — the candor is the point. (3) Whether a board of specialized agents decides better than its best single calibrated member; across seven regimes it did not — disconfirmed, and posted publicly. (4) Because it is a claim about Arclave graded by level with evidence and gaps shown, exactly the form the discipline demands of any assured decision. (5) Proven: capture, grounding, and sealing work on a controlled board. Promised: performance across a large live enterprise corpus, and the "catches what a lone expert misses" result, which is designed but not yet run.

Questions for reflection and discussion

When a vendor refuses to name what their product does *worst*, what should you infer — and what would you ask next?

The author pre-registered a hypothesis that turned out wrong and published it anyway. Where in your own work would a pre-registered prediction change how you judge a result later?

For facilitators: Put the seven-job scorecard on the screen and ask the group to grade a tool the organization currently uses, Built / Partial / Not yet. Watch which jobs draw disagreement — those are the jobs no one ever defined clearly enough to assess.

Which is more persuasive to you in a tool's favor — a flawless demo or a candid scorecard with two honest gaps? What does your answer say about how you currently buy?

Apply it

Pick one decision tool, platform, or "AI copilot" in your environment and build its seven-job scorecard: mark each job Built, Partial, or Not yet, write one line of evidence for each mark, and circle the two weakest. Then do the harder half — ask the tool's owner or vendor to grade it the same way, independently, and compare the two scorecards. *Trainers:* run the author's move in miniature — ask each participant to state, out loud, the single biggest gap in a tool they champion; the room learns more from those admissions than from any feature list, and it rehearses the candor the discipline runs on.

Notes and further reading

Not fooling yourself. Richard P. Feynman, "Cargo Cult Science," Caltech commencement address, 1974 (reprinted in *Surely You're Joking, Mr. Feynman!*, 1985): "The first principle is that you must not fool yourself — and you are the easiest person to fool."

The seven jobs. See the ideal-tool chapter ("The thing that catches it") for the capability set against which Arclave is graded here.

The disconfirmed board. The hypothesis that a multi-agent board out-decides its best single member was pre-registered on the Open Science Framework and disconfirmed across seven exploratory regimes; the results addendum is posted alongside the registration. (Pre-registration DOI 10.17605/OSF.IO/ZWM3S.)

Assurance, not control. See "Assurance vs. Control" for why a tool that does not out-decide the expert is still worth building — and why keeping the human in the chair is the design goal, not a limitation.

CONCLUSION

The Decisions of a Civilization

For twelve chapters we lived inside a single organization — its funding calls, its architecture sign-offs, its one bad Tuesday. Let me lift the lens now, all the way up, because the gap we set out to close does not stop at the company wall.

Think about the decisions that will shape this century. How a civilization meets a changing climate. How we govern the very artificial intelligence we just spent a book learning to distrust. How we hold a financial system that can seize in an afternoon and take a generation's savings with it. How we answer the next pandemic before it answers us. These are the largest decisions our species makes — and they are made, nearly all of them, the way that bank made its decisions: by serious people, with good intentions, real expertise, and a wall of dashboards, and with no system watching the whole, no grounding a stranger could check, no durable record, and no one whose single job is to assure. The blind spot is not corporate. It is civilizational. We instrument the particle and the stock tick to the millionth of a second, and we decide the future of the species on a confident slide and a nod in a hallway.

Half a century ago, two planners, Horst Rittel and Melvin Webber, gave this class of problem a name: *wicked* — not evil, but tangled past any clean solving. A wicked problem offers no stopping rule, no test for "correct," no second chance to run it again; every attempt changes the board. Climate is wicked. Pandemic response is wicked. Governing artificial intelligence is wicked. My own research began in that territory, looking for how anyone decides well when the problem itself will not hold still. And here is the uncomfortable truth I kept arriving at: when a problem resists solving, the *quality of the deciding* is the only thing left to hold onto. You cannot guarantee the outcome of a wicked problem. You can only make sure the decision was made completely, on the evidence, on the record, and answerable — so that when it turns, as it will, you learn from it instead of merely suffering it.

And now the accelerant. Everything that made the decision worth assuring at the scale of a company — the speed, the volume, the opacity, the machine quietly entering the loop — is arriving at the scale of a civilization, and faster. Artificial intelligence will draft the policy, price the risk, triage the ward, and model the climate; increasingly it will not just inform those calls but make pieces of them. The flood we met in Chapter 7 is rising over the whole world now, and the same hard symmetry holds: the technology multiplying the decisions is the only thing fast enough to assure them. We will not slow the flood. Anyone who promises to is not being straight with you. So we do the other thing. We build the ark.

Picture what that ark is, because it is not a machine that rules us. Three times before, humanity invented a quiet discipline that let it do something it could not do before. Double-entry bookkeeping let commerce grow past the reach of any single honest memory. The scientific method let knowledge accumulate instead of merely circulating. The independent audit let strangers trust a number, and capital markets the size of the planet grew on top of that trust. None of these solved anything by itself. Each one made a kind of *trust* possible at a scale that was impossible before — and on that trust, a civilization built.

Decision Assurance is that kind of discipline, aimed at the one thing the others left out: the decision itself. A world that assures its decisions — that makes them complete, grounds them in evidence, grades them by consequence, keeps them on the record, and lets an independent eye attest to them — is a world that can make its hardest collective choices and *learn from them*, out loud, together, instead of forgetting and repeating. That is not a world computer handing down verdicts. It is the opposite. It is the scaffolding that keeps the deciding human even as the deciding grows superhuman in speed.

There is a small sign that a discipline takes root: it becomes a verb. We no longer "perform quality assurance on" a part — we say it was assured. One day, I suspect, we will not say we "made the decision defensible." We will say we *arclaved* it — locked it so it holds, made it provable — and no one will need the longer phrase. You may be reading the word's first printed use. If the idea is right, it will not be the last.

Listen to the word, too. It stands a half-step from *archive* — but where an archive stores the past, to *arclave* is to lock the present so it can be proven. Do not only archive your decisions. Arclave them.

Let me hold the same honest line here that I held on every page, because a book that started in humility should not end in hubris. Assurance will not cool the planet or soften the next pandemic. It does not make a wicked problem tame, and it does not make a hard decision easy or a tragic one painless. It makes our decisions about those things *defensible, learnable, and ours.* And that last word — *ours* — is the whole of it. The danger of this age was never that the machines would decide badly. It is that we would hand them the deciding and forget that we did. Decision Assurance is the discipline that keeps humanity above the loop: the machine assures, and we decide, and we answer for it, all the way up to the decisions that belong to everyone. Reasonable, never absolute. In control, never replaced. That is the line between a tool of our freedom and a cage of our own making, and the entire purpose of this book is to keep us on the right side of it.

So I will end where I began, in a cafeteria, watching the cost of a decision no one could defend. I made a promise there, of two words. This book is my attempt to keep it — not for one company, but for all of us, because the gap I felt that afternoon is the same gap, only larger, behind every failure that ever made you wince at the news. The repair is not a product you buy and forget. It is a discipline we build together, in the open, practitioner to practitioner, owned by no one and in service of everyone — because that is the only way a thing like this lasts. I learned that word, *service*, at a Jesuit school a long way from here, and it turns out to be the most practical idea in the book. We assure decisions so that the people on the other side of them — your neighbors, your family, you — are dealt with well. Delight for all is not a slogan. It is the specification. And this book opens that work; it does not close it.

There is a line I keep returning to, and it is where the whole thing rests. *Somebody must audit the auditor.* At the scale of a company, that somebody is a profession we are only now naming. At the scale of a civilization, that somebody is us — together, awake, refusing to sleep on the job. We measured everything except the decision. Now we know how not to. So go and build the thing that catches it. Make the call you can prove was sound. And the next time a competent person asks you why you did what you did, keep the answer ready, and the evidence with it, and the quiet certainty that you did not decide in the dark.

Never again.

The book, in review

The lighter close, as the front and back matter carry.

What it all comes to

Decisions are the most expensive thing we make and the only one we never learned to assure — at every scale, from a Tuesday approval to the future of a species.

Decision Assurance is the discipline that makes a decision provably well-made: complete, consistent, governed, grounded, and on the record — the Quality Assurance of decisions.

It is **assurance, not control, and not an oracle**: it does not make the decision, and the experiment in this book found no machine that out-decides the expert in the chair. It makes the deciding sound and the human accountable.

Rigor matches consequence — the Decision Assurance Levels — so we assure the grave decisions fully and let the trivial ones flow.

A real discipline needs a **profession** — the decision auditor, a body of knowledge, a seal — and a **machine** that gives that profession reach without ever taking the human out of the loop.

The same forces are arriving at **civilizational scale**, where the decisions are wicked and the cost of getting them wrong is irreversible; assurance cannot tame a wicked problem, but it can make our choices about it defensible, learnable, and ours.

The point of all of it is **service** — that the people on the far side of every decision are dealt with well. *Somebody must audit the auditor. Let it be us.*

Two last terms

Wicked problem — a problem with no clean solution, no stopping rule, and no second run (Rittel & Webber, 1973);

at this scale, the quality of the deciding is the only thing you can hold onto.

Decision Assurance as civilizational infrastructure — a shared discipline, in the lineage of double-entry bookkeeping, the scientific method, and the independent audit, that makes a civilization's hardest collective decisions trustworthy without removing the human from the loop.

One last question

What is one decision — in your work, your institution, or your corner of the world — that you would want to *prove* you made well, ten years from now? Start there. Assure that one. Then the next.

APPENDIX A

The Decision Assurance Levels

Chapter 6 introduced the Decision Assurance Levels as a story — the monitor form and the rubber-stamped platform, the two axes, the single dial. This appendix turns that story into a specification you can lift onto a wall, encode in a workflow, or hand to a team on their first day. It restates the framework formally, defines each level in full, and lays out the rules that assign a level to a decision and the policy that levels thousands of decisions without a meeting. Nothing here contradicts the chapter; it simply renders the same idea at the resolution an implementer needs.

A note on status before we begin. This is an open specification, written to be owned by no single company and improved by the practitioners who use it. Treat the numbers and thresholds as defaults, not commandments: the structure is the contribution, and the structure is meant to be tuned to the risk appetite of the organization that adopts it.

A.1 The governing principle

The Decision Assurance Levels rest on one principle, borrowed wholesale from the most demanding safety-critical engineering humans practice: **rigor matches consequence.** A discipline that assures every decision to the same punishing degree buries its grave decisions under its trivial ones and grinds the organization to a halt; a discipline that assures nothing leaves its gravest calls naked. The answer the cockpit worked out — and that Chapter 5 traced through aviation's Design Assurance Levels — is to grade the rigor to the stakes, on a fixed scale, set by the consequence rather than by the convenience of the decider.

Everything in this appendix is machinery in service of that single sentence.

A.2 The two axes

A decision's level is set by two questions, asked in order.

Axis 1 — Severity. *If this decision turns out silently wrong, how bad is the result?* Severity is graded on a five-point scale, from existential to none:

Existential — the organization, or the people it serves, cannot recover. Bankruptcy, irreversible safety or environmental harm, a breach of a non-negotiable legal or ethical line.

Severe — a major, lasting wound that the organization survives but feels for years: a failed acquisition, a flagship product that collapses, a reputational injury that reshapes the market's trust.

Major — real money and real time lost, real disruption, but bounded and recoverable: a wrong vendor, a mis-set standard, a sizeable budget misallocated.

Minor — a manageable cost, absorbed in the ordinary course: a tactical misstep, a small overspend, a reworked plan.

None — no material effect on anything that matters: the cosmetic, the immaterial, the purely internal.

Axis 2 — Reversibility. *If this decision proves wrong, can you walk it back, and at what cost?* Reversibility is graded along a spectrum, but for working purposes it collapses to the distinction Chapter 6 borrowed from Jeff Bezos's 2015 framing:

Two-way door — reversible at low cost; you walk through, dislike what you find, and walk back out with little lost.

Heavy door — reversible, but only at significant cost in money, time, or disruption.

One-way door — effectively irreversible; once you pass through, the prior state cannot be restored at any acceptable cost.

Reversibility carries a hidden trap the specification names explicitly: a door that looks two-way on the contract can become one-way over time, as downstream decisions accumulate on top of it and lock-ins mature. The reversibility you assess must be the reversibility eighteen months out, not the reversibility on signing day. The Decision Flight Simulator's forward pass exists in large part to surface exactly this drift.

A.3 The worse-of-two rule

The level is set by the **worse of the two axes**, never by their average. A decision moderate in severity but irreversible is not a moderate decision; the door makes it grave. Formally:

> *A decision through a **one-way door** is never assigned below **Level B**, whatever its apparent severity. A decision of **existential** severity is **Level A**, whatever its reversibility.*

The worse-of-two rule is the single most important line in the specification, because it is the line that catches the decision the organization is most likely to wave through: the reversible-looking, modest-seeming call that quietly closes a door it can never reopen.

A.4 The five levels

Each level is defined by the decisions it fits and the assurance it earns. "What it earns" is the single dial of Chapter 6: set the level, and the level sets the depth of the five properties, the degree of independence, whether the simulator flies, whether a Decision

Audit and Seal apply, and whether the expected outcome is pre-registered.

Level A — Existential

Fits: a one-way door that could end the enterprise or cause irreversible harm — betting the company on a platform, an irreversible safety, compliance, or ethical call, a merger that cannot be unwound. Most organizations face only a handful a year, and that rarity is the point.

Earns: the full battery. All five properties (complete, consistent, governed, grounded, on the record) proven to the highest provable standard. An independent **Decision Audit** by someone outside the deciding line, ending in a **Seal of Assurance**. The **Decision Flight Simulator** flown forward to surface hidden dependencies and one-way doors before commitment. The decision and its expected outcome **pre-registered** — written down in advance — so the call can be learned from later without hindsight rewriting it. All dissent recorded verbatim. This is aviation's seventy-one objectives, translated into a boardroom.

Level B — Severe

Fits: a heavy door, costly and hard to reverse, that stops short of existential — a major strategic commitment, a senior hire, a market entry, a large sunk-cost program.

Earns: all five properties proven. **Independent review** — a second set of eyes the decider does not own and cannot overrule — short of the full audit apparatus. A forward run of the simulator on the gravest dependencies. A complete record, dissent included. A level-B decision should be sound enough that an auditor *would* seal it if asked, even when no formal audit is commissioned.

Level C — Major

Fits: the workhorse of real governance — a vendor selection, an architecture standard, a sizeable budget allocation, most of what an Architecture Review Board touches.

Earns: the five properties present and recorded, supported by a grounding-and-completeness **checklist**, with a home in the decision ledger. No formal audit rides on it, but the process behind it should be sound enough to survive one. Level C is where assurance earns its everyday keep: rigorous, but routine.

Level D — Routine

Fits: the two-way doors — most day-to-day approvals, tooling choices, small allocations, the reversible and the cheap-to-undo.

Earns: a light touch. Record the decision and the *why*, in a line or two, and otherwise trust the assured **process** around it rather than auditing the single instance. The discipline at Level D inverts: the failure mode is not under-assuring but over-assuring, and the correct move is to decide quickly, with good people, and move on. This is the door Chapter 6 warns against barricading.

Level E — Trivial

Fits: no real door at all — the cosmetic, the fully reversible, the immaterial.

Earns: nothing. Decide and move on. Rigor spent here is not virtue; it is the three-page form to requisition a second monitor. The discipline to assure *nothing* at Level E is as much a mark of maturity as the rigor lavished at Level A, because every minute of assurance spent on the trivial is a minute stolen from the grave.

A.5 The assignment rules

Four rules govern how a level is assigned. They exist to keep the framework honest under the two pressures that corrupt every grading scheme: the wish to talk a hard decision down, and the fear that drives everything up.

Rule 1 — The consequence assigns the level, not the decider. The level falls out of the two axes applied to the decision's actual consequence, the way a formal safety assessment sets a component's Design Assurance Level. No team may talk its own one-way door down to a two-way door because the scrutiny is inconvenient. Where the level demands independence, the assignment itself is made by someone other than the person who benefits from an easier grade.

Rule 2 — Take the worse of the two axes. As in §A.3. Severity and reversibility are assessed separately; the higher resulting level governs.

Rule 3 — When in doubt, default up. A decision that could plausibly be a B or a C is treated as a B until reversibility is shown to allow otherwise. The cost of over-assuring one decision is bounded and visible; the cost of under-assuring a grave one is unbounded and hidden. The asymmetry justifies the default.

Rule 4 — Re-level when the dependencies change. A level is not stamped once and forgotten. When new dependencies emerge — when the simulator's forward pass reveals that a "reversible" platform quietly becomes a one-way door in eighteen months — the decision is **re-leveled** upward, and the heavier treatment applies from that point. Re-leveling is a feature, not a failure: it is how the framework catches the door while it can still choose not to walk through.

A.6 Leveling as policy: governance-as-code

A framework that leveled every decision by a human, one at a time, would become its own three-page monitor form — the very overhead it exists to abolish. It does not. The leverage is in encoding the common cases as **standing rules** rather than per-decision **rulings**.

A *ruling* is a human weighing a single decision. A *rule* is a standing line that levels a whole class of decisions automatically, written once and applied thousands of times without a meeting. Examples of leveling rules an organization might adopt:

Any spend above a defined threshold, or any commitment that cannot be exited within ninety days, is at least Level B.

Any change to a customer-facing safety control, or any processing of a protected class of personal data, is Level A.

Any reversible, sub-threshold tooling choice is Level D — decide it and move.

Any cosmetic or internal-only change with no external effect is Level E.

Written once, those lines level the vast majority of an organization's decisions with no human deliberation, and reserve scarce human judgment for the two places it is most needed: the truly ambiguous, and the truly grave. This is **leveling as governance-as-code** — the leveling policy is itself a governed artifact: versioned, owned, reviewed, and assured like any other. And it inverts the overhead worry completely. A mature leveling policy makes the organization *faster*, because the two-way doors stop waiting in the one-way-door queue; the framework removes the gates that were applied by reflex and points the few that remain at the decisions that earn them.

A.7 Reference table

Level	Severity / door	The five properties	Independence	Flight Simulator	Audit + Seal	Pre-register outcome
A — Existential	existential, or any one-way door of grave severity	all five, fullest depth	full, independent Decision Audit	forward pass required	yes — Seal of Assurance	yes

B — Severe	severe; heavy or one-way door	all five, proven	independent reviewer (not full audit)	forward pass on key dependencies	sealable, not formally sealed	recommended
C — Major	major; recoverable at real cost	five present + checklisted	internal review	optional	none, but auditable	optional
D — Routine	minor; two-way door	light record of decision + why	none	none	none	no
E — Trivial	none; no real door	none	none	none	none	no

The table is the dial made legible: read across any row and you read the entire assurance posture a decision of that level earns.

A.8 Guardrails

Two cautions ride on the whole specification, in the spirit of the book.

A level is assurance *effort*, not a guarantee of outcome. Level A means a decision assured to the highest provable standard — *reasonable*, not absolute. Even aviation's seventy-one objectives never promised a perfect flight; they promised a defensible one. A Level-A decision can still meet bad luck and fail, and it remains sound. The level describes how hard the assurance worked, never how the dice fell.

Default up, but do not let fear pull everything to Level A. The framework earns its keep in both directions. It licenses grading *down* the reversible many — and that license is what makes it affordable to grade *up* the irreversible few. An organization that treats every decision as Level A did not achieve maximum safety; it rebuilt the hall of doors of Chapter 1 and called it diligence. The discipline is proportion, held against the two pressures that always push it out of true.

A.9 Notes

The graded-rigor principle and aviation's Design Assurance Levels (DO-178C) are established in Chapter 5; the five assurance properties in Chapter 4; the one-way / two-way door framing in Chapter 6 (Jeff Bezos, 2015 Amazon shareholder letter).

The Decision Flight Simulator (the forward pass that surfaces hidden irreversibility), the two ledgers (where the pre-registered expected outcome is later checked), the Decision Audit, and the Seal of Assurance are developed in Chapters 8 through 10; this appendix only specifies the dial that governs how much of each a decision earns.

This is version 0.1 of an open specification, offered for the Decision Assurance Body of Knowledge and meant to be revised by the practitioners who run it.

APPENDIX B

The Pre-Registration and Results

Chapter 3 told the human story of the experiment at the center of this book — the favorite idea that did not survive contact with the data. This appendix is the other half: the formal record. It states the hypothesis as it was registered, describes the method, reports the result across every regime tested, names the mechanism behind the failure, and draws the boundary of the claim with the precision the finding deserves. A reader who wants to argue with the conclusion should argue with this, not with the narrative — and that is exactly the point of writing it down.

One framing note before the record. The single most important property of this study is not its result but its *form*: the hypothesis, the measure, and the conditions that would refute it were committed to a public, timestamped registry **before** the experiment ran. That is what lets the disconfirmation mean something. A prediction you wrote down in advance and then watched fail is the most honest result a person can produce, because the writing-down removes the one move that lets a believer rescue a beloved idea after the fact.

B.1 The hypothesis

The study tested a claim the rest of the artificial-intelligence field treats as nearly self-evident, and that the author believed for twenty-five years: that a group, organized well, decides better than its best individual member. Applied to machines, the claim takes a sharp, testable form.

> ***H1 (the super-additivity, or "assembly-bonus," hypothesis).*** *A board of multiple artificial-intelligence agents, deliberating together, reaches decisions of higher quality than its single best, well-calibrated member acting alone.*

A natural rescue hypothesis followed, in case H1 failed for the obvious reason that the agents were too alike:

> ***H1a (the de-correlation hypothesis).*** *If a board of identical agents fails to beat its best member, then a board whose agents are deliberately de-correlated — drawn from different model families, with different training and different temperaments — will succeed, because diversity is the engine the wisdom of crowds is supposed to run on.*

Both hypotheses are claims of *super-additivity*: the assertion that the whole decides better than the sum, that an assembly bonus exists and can be captured by building the assembly. The intellectual lineage behind them is real and deep — the diversity-trumps-ability results, the wisdom-of-crowds literature, fifty years of meeting wicked problems with many minds rather than one. The study did not approach these as a skeptic. It approached them as a believer, looking for the proof.

B.2 Pre-registration

Before any data were collected, the hypothesis, the primary measure, and the conditions that would count as refutation were deposited on the Open Science Framework with a permanent identifier (DOI 10.17605/OSF.IO/ZWM3S). The deposit fixed, in advance and in public:

the claim under test (H1, and the rescue hypothesis H1a);

the operational measure of "decision quality" (agreement with an external expert verdict, defined below);

the comparison (a multi-agent board versus the single best agent on the identical inputs);

and the result that would refute the claim (the board failing to exceed its best member by the pre-specified margin).

The timestamp is the safeguard. With the criteria locked, no later analysis could quietly redefine success to match whatever the data showed.

B.3 Method

The design compared two configurations on the same decisions, scored the same way.

The decision task. The study used real accept-or-reject decisions of the kind expert panels render — including academic peer review, where each submitted paper already carried the verdict a human program committee actually reached. These human verdicts served as the *standard*: the yardstick against which both machine configurations were measured. (This role of the human panel is essential to the boundary of the claim, and §B.6 returns to it.)

The two configurations. Each decision ran two ways. First, through a **board** of artificial-intelligence agents — several models reasoning together, each assigned a distinct lens, deliberating to a collective verdict. Second, through the **single best** agent on its own. Both were scored identically: how often did each configuration's verdict match the human panel's actual call?

The seven regimes. The study did not run once. It ran across seven distinct regimes, because a single lucky or unlucky result settles nothing. The regimes varied the conditions a believer would say the assembly bonus needs:

a baseline board against its best member;

boards assembled from a single model family;

boards deliberately **de-correlated** across multiple model families (the direct test of H1a);

boards with an added **devil's advocate**, an agent whose only task was to argue against the emerging consensus;

boards governed by a **quorum** rule, requiring more than one agent to concur before a concern could sink a decision;

variations of domain and difficulty; and

combinations of the above, stacking every advantage the theory says the collective should need.

Across all seven, the comparison stayed constant: did the board beat its best single member at matching the expert standard?

B.4 Results

The result was consistent, and it was negative.

> ***Across all seven regimes, H1 and H1a were disconfirmed.*** *The board did not decide better than its best single, well-calibrated member. In several regimes it merely tied that member; in several others it did measurably worse — a panel of capable models reaching a poorer verdict than the strongest of them would reach alone. De-correlating the agents across model families did not rescue the result. Adding a devil's advocate did not rescue it. A stricter quorum rule made matters worse, by silencing the lone agent who was sometimes right. The assembly bonus the hypotheses predicted was not present in any configuration tested.*

The four attempted rescues — de-correlation, the devil's advocate, the quorum gate, and cross-family diversity — were each aimed at a different theory of why the board might be underperforming, and each failed on its own terms. After the fourth, the study stopped hunting for the configuration that would save the hypothesis and accepted the result.

B.5 The mechanism

A negative result is more trustworthy, and more useful, when the mechanism behind it is understood. Here it is.

A single, well-calibrated agent approves what deserves approval and rejects what deserves rejection at roughly the right rate — its accept-and-reject threshold sits in about the right place. Assemble a board of such agents and pool their concerns — the natural, sensible thing a board does — and the objections **accumulate**. Each agent flags something; the board honors all of it; and the collective starts rejecting decisions a calibrated individual would pass. The group does not average toward wisdom. It sums toward caution until its judgment tips.

The phrase for this is **miscalibration under aggregation**: the very act of combining many individually good judgments breaks the calibration each held on its own. Pooled carelessly, a group does not yield the best of everyone; it yields the sum of everyone's nerves. This is why the borderline cases — exactly the decisions where assurance matters most — are where the board failed hardest, drifting systematically toward over-rejection as each agent's separate worry was added to the pile.

B.6 The boundary of the claim

The finding is narrow, and the narrowness is its strength. The study compared a *collective* of artificial-intelligence agents to the *best single* artificial-intelligence agent. Both were machines. The human panel entered only as the measuring stick — the standard both configurations were scored against — never as a competitor in the contest.

Therefore:

The study makes **no claim** that a machine out-decides a human expert.

The study makes **no claim** that a human out-decides a machine.

That contest was never staged, and nothing here speaks to it.

What the study tested, and all it tested, is whether *assembling* artificial-intelligence agents into a board makes them decide better than the best one alone. The answer, across seven regimes, was no.

Two further limits belong on the record. First, the result is **exploratory**, drawn from public and synthetic decision corpora rather than from a large body of consented, real-world enterprise decisions. A separate, **confirmatory** study — pre-registered in draft, built around a novel-catch metric on real enterprise cases — is designed and not yet run; until it is, the finding should be read as strong and directional, not as the final word. Second, the result concerns decision *quality* (agreement with the expert verdict), not the many other things a board of agents does — and one of those other things turned out to matter more than the verdict, which is the subject of the next section.

B.7 The discovery in the wreckage

Buried in the negative result was the finding that turned a dead end into the thesis of this book.

When the agents erred, they often erred *together* — confidently, in chorus, fabricating the same kind of unfounded objection, so that adding more agents amplified the error rather than canceling it. But when each agent was **forced to ground every claim** — to point at the actual text and quote it, or else stay silent — the fabrications dropped by roughly half. The fix for a confident machine error was not more machines voting. It was grounding: making each claim show its evidence or withhold itself.

That single result reframed everything. The board was a poor *oracle* — it could not vote its way to a better verdict — but along the way it did something else flawlessly: it tracked every concern, applied every rule the same way every time, and produced a complete, inspectable record of how the decision was reached. It was a magnificent *assurance* system. The study set out to build a machine that decides better and discovered, in its own disconfirmed data, a machine that makes deciding *provable*. The dream of the smarter machine died here. Decision Assurance is what rose in its place.

B.8 What it means for the book

This appendix is the empirical floor under the entire argument. The book does not claim that an assembly of agents will out-decide your experts, because the author tested that claim, pre-registered it, and watched it fail across seven regimes. The positioning of the whole work — *assurance, not an oracle; the human keeps decision control while the machine makes the decision provable* — is not a marketing posture chosen for modesty. It is the only conclusion the evidence permits. A discipline that began in a disconfirmed hypothesis carries a humility that a discipline born in a confident pitch never can, and that humility is the source of its credibility, not a limit on it.

B.9 Notes

The parent pre-registration is public on the Open Science Framework (DOI 10.17605/OSF.IO/ZWM3S). A dated **Results Addendum** reconciling the registered hypothesis against the observed outcome is posted alongside it; a cross-regime **Capstone** synthesis collects the seven regimes; a **Tier-2 confirmatory pre-registration** (a novel-catch metric on consented real-world cases) is drafted and, as of this writing, not yet posted.

The human story of this study — the conviction, the loneliness of publishing the evidence against one's own case, and the grounding discovery — is Chapter 3. This appendix is its evidentiary companion; figures and exact regime details live in the OSF deposit and should be cited from there before print.

"Wisdom of crowds," "diversity trumps ability," and the wicked-problems lineage that motivated H1 are discussed in Chapter 3 (Surowiecki; Page; Rittel & Webber).

APPENDIX C

A Decision, Worked End to End

The chapters take the discipline apart to examine each piece. This appendix puts it back together. It follows one decision — a hard, realistic one — from the moment it lands on the table to the moment, a year later, when it teaches the next decision something. The seven jobs of an assurance tool, scattered across the framework chapters for teaching, here run in sequence on a single call, the way they would on a real Tuesday. Read it as the assembled machine, turning.

The names and numbers are illustrative. The shape is not.

The decision

The Architecture Review Board (ARB) receives a request. A business unit wants to deploy an Artificial Intelligence (AI) customer-service agent — not a chatbot that answers questions, but an agent with authority to *act* on customer accounts: issue refunds, change service plans, and, in defined cases, close accounts. The vendor's pitch is strong: the agent resolves routine requests in seconds, and the unit projects savings of roughly thirty full-time equivalents (FTEs). The unit wants approval to roll it out across all customer accounts within ninety days.

You chair the board. The old way, you could approve it on the strength of the demo and the savings number, or reject it on instinct — and either way leave almost no trace of why. The assured way runs differently. Watch each job do its part.

Job 1 — Capture: make the decision complete

The first job is completeness: make sure nothing that matters is missing before the board even debates. The tool captures the decision in a structured frame rather than a free-text ticket titled "AI agent — approve?"

The question, stated precisely. Grant an autonomous AI agent authority to take irreversible actions on live customer accounts, across the full customer base, within ninety days.

The options, more than two. Approve as requested; approve with the agent restricted to *reversible* actions only; approve a limited pilot on one low-risk account segment; reject. Four options on the table, not the false binary of yes-or-no the request arrived as.

The stakeholders who must be heard. The business unit, yes — and also security, privacy, legal, the complaint-handling team, and the frontline staff whose work the agent would absorb.

The evidence the board will need. The vendor's accuracy claims, the savings model, the regulatory constraints on automated account actions, and the reversibility of each action type.

Capture alone changes the conversation. The request walked in as "approve the AI agent." It now sits on the table as a four-option decision with named stakeholders and a defined evidence list. Half of bad decisions are bad because the framing was too narrow to begin with; this job widens it before any debate starts.

Job 2 — Ground: let no claim float

Now the tool grounds every claim — each assertion must point at real evidence or be marked unverified. Three claims anchor the request, and grounding tests all three.

The vendor states "97 percent accuracy." Grounded, the number tells a different story: it traces to a benchmark the vendor ran on clean, well-formed requests — not to anything resembling this company's messy, real-world account data. The claim is true *and* irrelevant, and grounding is what surfaces the difference. The board does not get to treat 97 percent as a fact about its own customers, because the evidence does not reach that far.

The "thirty FTEs saved" traces to a model that assumes the agent handles eighty percent of ticket volume unaided — itself resting on the same optimistic accuracy figure. One ungrounded number was quietly holding up two others.

The regulatory claim — "automated account actions are permitted" — pulls back, when grounded, a constraint the request omitted: closing a customer account through a fully automated process triggers a notice-and-appeal obligation in two of the company's operating regions. Not a blocker. A condition the request never mentioned.

Grounding did not kill the decision. It removed the fog around it, so the board now debates what is real rather than what was confidently asserted.

Job 3 — Level: match the rigor to the stakes

How much assurance does this decision earn? The Decision Assurance Levels (DALs) answer with two axes and the worse-of-two rule.

Severity: a wrong action on a single account is severe but bounded — a refund clawed back, a plan restored. At the scale of the full customer base, a systematic error becomes a different animal: thousands of wrongful closures, a regulatory finding, a public breach of trust. Severity rates **severe, trending existential at scale.**

Reversibility: a refund is a two-way door. A *wrongful account closure* — the relationship severed, the customer gone, the trust spent — is, for that customer, a one-way door.

Worse-of-two governs. A decision that opens a one-way door at scale is **Level A**, whatever the savings promise. The leveling is not a matter of the board's appetite; it falls out of the consequence. And Level A earns the full battery: all five properties proven, an independent audit ending in a seal, the flight simulator flown forward, the expected outcome pre-registered, every dissent recorded. The savings number does not get to talk the level down.

Job 4 — Map and simulate: rehearse before committing

The tool places the decision on the Decision Dependency Graph and flies it.

The graph shows what the agent would touch: the refund policy, the identity-verification system, the account-closure workflow, the complaint pipeline, the audit-log store. One node lights up as the structural risk — the agent's authority to take *irreversible* actions without a human in the path.

The forward pass confirms it: if anything fails, the closures are where the damage concentrates and cannot be undone. The outcome pass, run across many sampled futures, prints a spread — in most runs the agent performs well, but a real tail produces clusters of wrongful closures, precisely in the messy-data conditions the grounding step already flagged. The backward pass, starting from "safe go-live, no wrongful irreversible action," traces the route home: irreversible actions must sit behind a human gate, and every automated action must carry a reversibility window before it becomes final.

The simulator did not predict the rollout. It rehearsed it — and every mode pointed at the same node the grounding step distrusted. Two independent jobs, the same finger, the same risk.

Job 5 — Audit and seal: an independent set of eyes

Because the decision is Level A, it earns an independent **Decision Audit** — a reviewer outside the deciding line, who benefits from no easier grade. The auditor does not re-decide; the auditor checks the *assurance*. Was the decision complete? Grounded? Leveled correctly? Were the conditions sound and the dissent recorded?

The audit finds one gap and one strength. The gap: the original savings case still leans on the vendor's clean-data accuracy, which grounding already discredited — so the auditor requires the projection be rebuilt on a conservative, this-company estimate before the number enters the record. The strength: the human-gate condition the simulator demanded is present and well specified. With the savings claim corrected, the auditor issues the **Seal of Assurance** — a record not that the decision is *right*, but that it was *made well*: everything that should be checked was checked.

Job 6 — Record: seal the snapshot

The decision is captured in the **Assurance Record** and sealed — the immutable snapshot of the call as it was made:

The question and the four options, with the costed case for each.

The grounded evidence, including the explicit note that the vendor's 97 percent does not transfer to this company's data.

The level: A, by worse-of-two, with the reasoning.

The conditions: a human gate on all irreversible actions; a ninety-day reversibility window on automated actions; a monthly accuracy audit on *this company's* live data; notice-and-appeal honored in the two regions that require it.

The dissent: the security lead argued even reversible actions deserve a slower rollout; overruled, with the reason recorded beside the objection, not over it.

The owner: a named human, accountable for the call.

The expected outcome, pre-registered: deflect forty percent of low-risk tickets within two quarters; zero wrongful irreversible actions; customer satisfaction (CSAT) steady within two points.

The board approves option two-and-a-half — a pilot that expands on evidence, with the human gate permanent. The record is sealed. Whatever happens next, the reasoning is now retrievable, not reconstructable.

Job 7 — Remember: let the outcome teach without rewriting the past

Twelve months on, the **Outcome Ledger** records what actually happened. Deflection reached thirty-five percent, not forty. The human gate caught and reversed eleven wrongful-closure attempts — the exact tail the simulator predicted, stopped before a single customer was lost. CSAT held steady. The agent's measured accuracy on the company's real data ran several points below the vendor's benchmark, precisely as the grounding step warned a year earlier.

Now the one rule does its work. A one-ledger organization would read "thirty-five percent, not forty" as a miss and punish the sponsor. The two-ledger organization opens the sealed record, sees that the decision was complete, grounded, correctly leveled, and properly conditioned — and that its conditions *caught the only serious failure mode before it cost anything* — and leaves the grade exactly where it is: well-made. The shortfall does not re-grade the past. It *calibrates the future:* the delta between the forty percent expected and the thirty-five percent delivered becomes a standing lesson — *never let a vendor's clean-data benchmark stand as grounding; re-test on our own data before we believe a number.* That lesson lands on the next AI-agent proposal as a rule at the gate, written once and applied to every proposal after.

The assembled machine

Step back and look at what the seven jobs did, together, that no one of them could do alone. Capture widened a yes-or-no into a real choice. Grounding stripped out a number that would otherwise justify far too much. Leveling matched the rigor to a one-way door at scale. The map and the simulator rehearsed the one failure that mattered before it could happen. The audit caught the discredited claim trying to sneak back in, and the seal put an independent name on the rigor. The record made the whole thing retrievable. And memory turned a single assured decision into a lesson the institution keeps.

Notice who decided. The board decided. A human owns the call, and a human owns the eleven reversals the gate routed back for review. The tool never took the controls. It made the decision complete, honest, proportionate, rehearsed, independently checked, recorded, and — a year later — instructive. That is the difference between a decision you hope was right and a decision you can prove was made well. It is the whole discipline, turning on one call.

THE FIELD KIT

The guard — read it first. A checklist prompts a decision; it never replaces one. It assures that you thought; it does not think for you. Tick a box only when you can show the thing behind it. A checklist run on autopilot is the rubber stamp this book warns against — the form filled, the judgment skipped.

Checklist 1 — The Decision Assurance Scorecard

Source: Chapter 4 (Assurance versus Control). Grade one decision type — a funding approval, an architecture sign-off, a credit call — property by property. Score each line provable, partial, *or* absent. *The gaps are your assurance backlog.*

Complete — Is there a defined list of what *must* be weighed for a decision of this kind, and can you show each item was, in fact, weighed? *(If the list itself does not exist, completeness is not low — it is unmeasured.)*

Consistent — Take two similar decisions from different months and different people: did the same visible rule produce both? *(If they diverge and no one can point to why in the criteria, circumstance decided, not standard.)*

Governed — Name the constraints that must never break. What in the system — not someone's memory — would actually stop a senior person from overriding one? *(If the answer is "nothing but goodwill," the rule is a suggestion in a uniform.)*

Grounded — Take the central claim the decision rested on and follow it to a source a stranger could check. *(If the trail ends at an assertion, the decision was asserted, not grounded.)*

On the record — Pick a consequential decision from a year ago and reconstruct *why* from what was written down at the time. *(How far you get before the trail goes cold is the measure of your institutional memory.)*

A decision that scores provable *on all five is* assured. *One or two is the usual starting score. The rest is the work.*

Checklist 2 — Pre-flight: "Assure this decision"

Source: Chapter 11 (Monday morning). The before-you-commit pass for a single high-stakes call — the seven-step sequence, compressed to a pre-flight card. Run it on the one-way doors, not the routine many.

Pick. Is this a one-way door? *(If it is reversible and cheap to undo, assure it lightly and move on; save this pass for the doors that do not reopen.)*

Capture. Write the question, the options weighed, the evidence in hand, and the owner's name — in minutes. *(If capturing costs more than deciding, no one will; keep it light.)*

Ground. Tie every load-bearing claim to a source a stranger could check. *(If the trail ends at an assertion, it is not grounded yet.)*

Level. Severity by reversibility, worse of the two. *(The level sets how much of the rest you owe — see Checklist 4.)*

Rehearse. For a Level A or B call, run the dependencies forward: what does this shake downstream, and where does a "reversible" door quietly close eighteen months out?

Record and seal. Assemble the above into an Assurance Record and seal it, with the expected outcome written down. *(This is the answer to the audit email, eight months early.)*

Decide. Make the call — owner named, dissent recorded. The assurance does not decide; you do.

Checklist 3 — Vendor / tool evaluation

Source: Chapter 2 (the seduction of the smarter machine) and Chapter 10 (the ideal tool). Grade any decision tool — an "AI decision platform," a copilot, an agentic board — against the seven jobs before you trust it. Mark each Built, Partial, *or* Not yet, *and make the vendor name the two it does worst.*

Capture — Does it record the question, options, evidence, and owner in the flow, or only store an answer?

Ground — Does it cite checkable evidence for its claims, and abstain when it cannot — *including for its own assertions?*

See the whole — Can it show a decision's dependencies (the graph), or is that still a slide?

Level — Does it grade rigor to consequence, or treat every decision the same?

Simulate — Can it rehearse a decision forward *before* you commit, or only react after?

Assure and seal — Does it produce a sealed record you could hand an auditor, or just print a confident verdict?

Remember — Does it keep the two ledgers (decision and outcome) and calibrate *without* re-grading the past?

The disqualifier (Chapter 2): does it keep the human deciding, or quietly take the controls? A tool that claims all seven, fully, today — no Partial, *no* Not yet *— is selling the oracle this book refuses. Thank them, and leave.*

Checklist 4 — Decision Assurance Level: match rigor to consequence

Source: Chapter 6 (How much assurance is enough?). It does for a decision what aviation's Design Assurance Level does for flight software: it sets how much assurance a decision deserves by the size of the consequence. Run it before you spend — or skip — the rigor.

If this turns out silently wrong, how bad is it? Existential (you cannot recover), severe, major, minor, or no real effect?

Is it reversible? A two-way door you can walk back through needs less assurance than a one-way door.

Take the worse of the two. An irreversible door is never below Level B, whatever the severity feels like in the moment.

When you cannot tell, default up one level — then ask whether the reversible truth lets you settle back down.

Set the dial. The level you land on sets everything else: how much of the five properties to prove, how much independence to demand, whether the Flight Simulator flies, and whether a Decision Audit and Seal apply.

Most teams find the same mismatch: heavy assurance on small reversible calls, light assurance on large one-way doors. Naming the level is how you re-allocate.

Each checklist above is the canonical version; the in-chapter copy points here. Use them together: Checklist 4 sets a decision's level, Checklist 2 walks a high-stakes one through assurance, Checklist 1 grades a decision type after the fact, and Checklist 3 judges the tools that promise to help.

GLOSSARY

This glossary is **normative**, not decorative. The terms below are the canonical first definitions of the Decision Assurance discipline, set down here so the field can share a vocabulary from its first day. They are offered as the seed of the **Decision Assurance Body of Knowledge (DABOK)**, to be inherited, governed, and extended by the **Decision Assurance Council** — versioned like any other governed artifact, exactly as the discipline itself prescribes. Where a definition restates a term established elsewhere (audit, the wisdom of crowds, the digital twin), the source is named in the References. Where a term is new to this work, the definition here is the reference one.

Entries are alphabetical. A *see also* points to the related terms that complete the picture.

Accountability (versus authorship). Answerability for a decision — the obligation to stand behind it and bear its consequences. Accountability rests always on a human or an organization, and never transfers to a machine, even when the machine performs the authorship (the drafting, ranking, or scoring). The distinction is the moral spine of the discipline: a model may help make a decision; only a person can answer for it. *See also* Decision Assurance; Accountability vacuum.

Accountability vacuum. The gap that opens when an autonomous system makes or shapes a consequential decision and no human is positioned to answer for it. Decision Assurance exists in large part to close this vacuum by keeping a named human owner on every assured decision. *See also* Accountability (versus authorship).

Append a correction. The discipline that keeps an immutable record honest: a genuine error in a sealed record is set right by a dated, signed addition recorded *alongside* the original, never by silently editing the original. Immutability means uneditable, not uncorrectable. *See also* Assurance Record.

Arclave. The author's Decision Assurance tool — an Architecture Review Board (ARB) digital twin built to perform the assurance jobs around a real decision while the human keeps the decision. Arclave is one implementation of the discipline, graded in public against the seven jobs in the book's penultimate chapter; the discipline does not depend on it. *See also* The seven jobs; The honest scorecard.

arclave (verb) /ˈɑːrk.leɪv/ — to lock a decision so it holds; to make it provable. The discipline's name as an everyday act: to *arclave* a decision is to assure it — capture, ground, level, seal, and record it — so it holds up afterward. Coined in this work (see the verb register in "How to cite this work"). *Example:* "We arclaved the cloud-migration call before sign-off." *See also* Decision Assurance; Assurance Record; Seal of Assurance.

Assembly bonus. *See* Super-additivity.

Assurance Record. The immutable, sealed, per-decision ledger: the complete snapshot of a single decision at the instant it was made — the question, the options weighed, the evidence and where it was grounded, the assurance level, the conditions attached, the dissents recorded, the named human owner, and the pre-registered expected outcome. It is the retrievable answer to the audit email. *See also* Outcome Ledger; Calibrate, never re-grade; Seal of Assurance.

Backcasting. A planning method that starts from a desired future and works backward to the decisions that would reach it, inverting the usual forward question; one of the three modes of the

Decision Flight Simulator. Coined by John B. Robinson (1982). *See also* Decision Flight Simulator.

The boundary of the claim. The explicit limit on what the book's pre-registered experiment tested: a board of artificial-intelligence agents was compared to the single best artificial-intelligence agent — machine against machine, never machine against human. No claim is made, anywhere, that a machine out-decides or under-decides a human expert. *See also* Pre-registration; Super-additivity.

Calibrate, never re-grade. The governing rule of decision memory: outcomes are allowed to *calibrate* the system going forward (sharpening levels, checklists, and forecasts) but are never allowed to reach back and *re-grade* a past decision. A decision well-made on the evidence available stays well-made, even when the result is poor. The structural cure for resulting. *See also* Resulting; Outcome Ledger; Sound decision.

Certified Decision Auditor (CDA). The credential of the decision-auditing profession: examined competence, a code of ethics, continuing study, and the standing to place a Seal of Assurance. The decision-world counterpart to the Certified Public Accountant and the Certified Internal Auditor. *See also* Decision auditor; Seal of Assurance.

Decision Assurance. The discipline of systematically ensuring that a decision is made well — completely, consistently, within its governing rules, and on real evidence — and recording it so the decision can be proven and defended afterward, with accountability always resting on a human or an organization, whether the decision is made by a person, an artificial intelligence, or the two together. The Quality Assurance of decisions; the proactive counterpart to Decision Control. *See also* Decision Control; The five assurance properties; Sound decision.

Decision Assurance Body of Knowledge (DABOK). The shared, open, teachable account of the discipline — its definitions, principles, methods, and standards of practice — that turns a private knack into a public profession. This book is offered as its seed; the Decision Assurance Council is its proposed steward. *See also* Decision Assurance Council; Decision auditor.

Decision Assurance Council. The proposed governing body of the discipline: the steward of the Decision Assurance Body of Knowledge, the Decision Assurance Levels specification, the auditor credential, and the canonical lexicon — versioning and governing them as the discipline matures. *See also* Decision Assurance Body of Knowledge.

Decision Assurance Level (DAL). The graded rigor a decision earns, set by its consequence rather than the convenience of the decider; five levels, A (existential) through E (trivial), adapted from aviation's Design Assurance Level. The level is the single dial that sets every other safeguard at once. *See also* Severity; Reversibility; The dial; Design Assurance Level.

Decision Audit. The engagement in which an independent decision auditor assures a decision *process* — or, at Level A, a single decision — against the five properties at the decision's level, and returns a grade plus a list of gaps. An audit of the *making*, never a verdict on the future outcome. *See also* Decision auditor; Seal of Assurance.

Decision auditor. An independent professional who assures the *making* of a decision — against the five properties, at the decision's level — without making the decision. The decision-world counterpart to the financial auditor; the standing answer to "who audits the auditor?" *See also* Certified Decision Auditor; Independence; Decision Audit.

Decision Control. The reactive discipline of judging an individual decision — ratifying or rejecting it at the gate, and monitoring it thereafter — so a flawed decision is caught at or near the moment it is made. The Quality Control of decisions; the established term in agency theory (Fama and Jensen, 1983). The decision-by-decision counterpart to Decision Assurance. *See also* Decision Assurance.

Decision debt. The compounding cost of decisions made with no record, no rationale, and no way to learn from them; the quiet, costlier cousin of technical debt. Like all debt left in the dark, it accrues silent interest and comes due later. *See also* Decision ledger; Institutional amnesia.

Decision Dependency Graph. A model of decisions as nodes connected by directed dependencies. Reading the arrows one way reveals a decision's *antecedents* (everything it rests on); reading them the other way reveals its *dependents* (everything resting on it). The overdue cousin of the critical-path networks project managers built in the 1950s. *See also* Digital twin; Decision Flight Simulator.

Decision fatigue. The depletion that comes from making too many choices, so judgment degrades as the day wears on — and the biggest calls often land precisely when the depletion is deepest. *See also* Decision Assurance.

Decision Flight Simulator. The what-if capability run on the digital twin of the decision system, in three modes — **forward** (what does this decision do across the graph?), **outcome** (how might it go, sampled across many runs?), and **backward** (what decisions reach the future we want?). It *rehearses* a decision's consequences rather than predicting them; the outputs inform the human, who still decides. *See also* Decision Dependency Graph; Digital twin; Rehearse, do not predict.

Decision ledger. A running record of decisions, captured as they occur, the way a bookkeeper records transactions in a general ledger; the humble first practice from which the discipline grows, and the everyday ancestor of the Assurance Record. *See also* Assurance Record; Decision debt.

Default-up rule. When a decision's level is unclear, assign the higher one, then grade back down only if reversibility allows. The cost of over-assuring one decision is bounded and visible; the cost of under-assuring a grave one is unbounded and hidden, and the asymmetry justifies the default. *See also* Decision Assurance Level; Re-leveling.

Design Assurance Level. In aviation's DO-178C standard, the level of rigor assigned to a software component by the severity of its failure, from A (catastrophic) through E (no effect); the direct ancestor of the Decision Assurance Level. *See also* DO-178C; Decision Assurance Level.

The dial. The principle that a decision's level sets every other safeguard at once — the depth of the five properties, the degree of independence, whether the simulator flies, whether an audit and seal apply, whether the outcome is pre-registered — rather than each being negotiated separately. *See also* Decision Assurance Level.

Digital twin. A virtual mirror of the real decision system, kept in sync with it and accurate enough to experiment on safely; the world the Decision Flight Simulator flies. Borrowed from engineering (Grieves, circa 2002; named at NASA, circa 2010), with a conceptual ancestor in NASA's paired Apollo-era ground models. *See also* Decision Dependency Graph; Decision Flight Simulator.

DO-178C. *Software Considerations in Airborne Systems and Equipment Certification* — the standard governing airborne software, which grades each component by the consequence of its failure (software levels A through E) and is the engineering template for graded decision rigor. *See also* Design Assurance Level.

Emergent Intelligence. The author's parent theory, of which Decision Assurance is the first applied discipline: the view of artificial intelligence through the lens of complex adaptive systems rather than linear reductionism, in which simple rules applied by many autonomous agents produce capability greater than the sum of the parts. The pattern that emerges from a stream of assured decisions is, in this frame, the shape of intelligence itself. *See also* Decision Assurance; Calibrate, never re-grade.

The five assurance properties. The five qualities that together make a decision *sound*: **complete** (nothing decisive missing), **consistent** (the rules applied the same way every time), **governed** (within the rules that must not break), **grounded** (resting on checkable evidence), and **on the record** (captured so it can be proven later). The functional test of Decision Assurance. *See also* Decision Assurance; Sound decision; Grounding.

Governance as code. Expressing the assurance rules — levels, required evidence, sign-offs — as policy the system enforces automatically and consistently, rather than guidance a reviewer is trusted to remember; the broader cousin of "policy as code." Applied to leveling, it lets most decisions self-level and reserves human judgment for the ambiguous and the grave. *See also* Leveling policy; Decision Assurance Level.

Grounding. The requirement that a claim point at checkable evidence — quote the text, cite the source — or else abstain. In the book's pre-registered research, forcing agents to ground every claim roughly halved their fabrications; grounding is the in-tool answer to "who assures the assurer?" *See also* The five assurance properties; Hallucination.

Hallucination. A confident, fluent, fabricated output — a fact or a rationale — presented as if true. The characteristic failure of generative artificial intelligence, and the reason grounding is load-bearing. *See also* Grounding.

The honest scorecard. A job-by-job statement of what a Decision Assurance tool does well today and where it still falls short, marked plainly (for example: Built, Partial, Not yet). Applied in the book to the author's own tool, graded in public; the gaps cluster exactly where human judgment, independence, and trust stay essential. *See also* The seven jobs; Arclave.

Independence. The assurance principle that the checker must not be the maker — the auditor who does not work for the decider, the DO-178C objectives met "with independence." The standing answer to "who checks the checker?" *See also* Decision auditor; Reasonable assurance.

Information Assurance. The discipline of making information's soundness provable; the United States Department of Defense names five pillars — confidentiality, integrity, availability, authentication, non-repudiation — extending the older confidentiality-integrity-availability triad. One of the assurance lineages Decision Assurance inherits. *See also* Quality Assurance / Quality Control.

Institutional amnesia. An organization's loss of its own hard-won lessons as the people who held them move on; the condition in which the same debate reopens for the third time with none of the first two recorded. The Outcome Ledger and the Assurance Record exist to end it. *See also* Decision debt; Outcome Ledger.

The irony of automation. The finding that automating most of a process leaves the remaining human a harder, more critical, monitoring-and-rescue role, not an easier one (Lisbeth Bainbridge, 1983); a caution against believing a machine removes the need for human judgment. *See also* Accountability vacuum.

Leveling policy. The set of standing rules that assign decision levels by class rather than one decision at a time — written once and applied thousands of times without a meeting — so most decisions self-level and scarce human judgment is reserved for the ambiguous and the grave. A governed, versioned artifact in its own right. *See also* Governance as code; Decision Assurance Level.

Limited assurance. *See* Reasonable versus limited assurance.

LLM-as-judge. Using a large language model (LLM) to grade or rule on outputs; in this framework, Decision Control automated — useful, but not a substitute for assurance, and subject to its own biases. *See also* Decision Control; Hallucination.

Minimum viable practice. The smallest real implementation of Decision Assurance — one consequential decision assured completely — used to prove and grow the discipline rather than launch it all at once; the governance adaptation of the minimum viable product. *See also* Governance as code.

Miscalibration under aggregation. The failure mode in which pooling many individually well-calibrated agents' concerns degrades the calibration each held alone, producing systematic over-rejection; the mechanism behind the book's disconfirmed assembly-bonus hypothesis. *See also* Super-additivity; The boundary of the claim.

On the record. *See* The five assurance properties.

One-way / two-way door. *See* Reversibility.

Outcome Ledger. The cross-decision ledger of what actually happened over time, decision after decision; the source of calibration, kept in a separate book from the Assurance Record by design so that a later outcome can never rewrite a past decision's grade. *See also* Assurance Record; Calibrate, never re-grade; Reference class.

Pre-registration. Committing a hypothesis, method, and the criteria that would refute it to a public, timestamped record *before* running a study, so results cannot be quietly reinterpreted after the fact. The safeguard behind the book's disconfirmed-board finding (Open Science Framework, DOI 10.17605/OSF.IO/ZWM3S); at Level A, decisions are pre-registered for the same reason. *See also* The boundary of the claim; Decision Assurance Level.

Prospective corpus. Assurance demonstrated in a controlled, instrumented setting rather than across a large body of live decisions; a claim still being earned rather than settled. Used in the book to mark, plainly, the present limit of the author's evidence. *See also* The honest scorecard.

Quality Assurance / Quality Control. The manufacturing distinction the whole discipline is modeled on: building the process so defects cannot form (assurance) versus inspecting the finished item to catch the defect that did (control) — "build quality in, do not inspect it in" (Deming). Decision Assurance is the first; Decision Control is the second. *See also* Decision Assurance; Decision Control.

RAID log. A project register of Risks, Assumptions, Issues, and Decisions; the familiar, well-meant, per-project record whose limits (it dies with the project, never aggregates, rarely keeps its context) the enterprise decision ledger exists to overcome. *See also* Decision ledger; Retrospective.

Reasonable assurance. In audit, a high but deliberately *not absolute* level of assurance — the honest admission that no real process drives risk to zero. Decision Assurance inherits the humility: a Level-A decision is assured to the highest provable standard, which is reasonable, never perfect. *See also* Reasonable versus limited assurance; Independence.

Reasonable versus limited assurance. The auditor's two honest registers of confidence: *reasonable* (high but not absolute, the standard register) and *limited* (the lighter form, used when the evidence is thinner). A Seal of Assurance states which register it carries. *See also* Seal of Assurance; Reasonable assurance.

Reference class (the outside view). The set of similar past decisions whose base rate predicts the present one better than its own seductive details do (Kahneman and Tversky; reference-class forecasting, Flyvbjerg). The Outcome Ledger is what finally makes an organization's own reference class possible. *See also* Outcome Ledger; Calibrate, never re-grade.

Rehearse, do not predict. The discipline's stance on foresight: a simulator's value is the readiness its rehearsal builds, not a forecast it claims. The Decision Flight Simulator informs the human, who still decides; a tool that claimed to know the future would be the oracle the book refuses. *See also* Decision Flight Simulator; Sound decision.

Re-leveling. Revising a decision's level when its dependencies reveal hidden irreversibility — for example, when the simulator's forward pass shows a "reversible" platform quietly becoming a one-way door in eighteen months. A feature, not a failure: it catches the door while there is still a choice not to walk through it. *See also* Decision Assurance Level; Default-up rule.

Resulting. The error of judging a decision's quality by the quality of its outcome (Annie Duke, borrowing the word from poker); the most expensive memory error an organization makes, because it punishes the unlucky and rewards the fortunate and teaches everyone the wrong lesson. *See also* Calibrate, never re-grade; Sound decision.

Retrospective. The agile practice of reviewing what worked and what did not, to improve the next iteration; a real instinct toward decision memory that, like the RAID log, lacks a home that outlives the project. *See also* RAID log; Outcome Ledger.

Reversibility (one-way / two-way door). Whether a decision can be walked back, and at what cost: a **two-way door** is reversible at low cost; a **one-way door** is effectively irreversible (Jeff Bezos, 2015). One axis of a decision's level; a one-way door raises the level regardless of apparent severity. The reversibility that counts is the reversibility eighteen months out, after lock-ins mature — not on signing day. *See also* Severity; Decision Assurance Level; Re-leveling.

Seal of Assurance. The mark an accredited decision auditor places on a decision or process, stating the register of confidence (reasonable or limited) and the Decision Assurance Level met; an audit opinion on the *making* of the decision, never a prophecy about its outcome. *See also* Decision Audit; Certified Decision Auditor.

Severity. How damaging it is if a decision turns out silently wrong, graded from existential to no effect; one of the two axes that set a decision's level, taken together with reversibility by the worse-of-two rule. *See also* Reversibility; Decision Assurance Level; Worse-of-two rule.

The seven jobs. The functional spine of a Decision Assurance system: **capture** (make the decision complete), **ground** (let no claim float), **see the whole** (the dependency graph), **grade the level** (the Decision Assurance Level), **simulate** (the Flight Simulator), **assure and seal** (the audit and the Seal), and **remember** (the two ledgers). Turned into buyer's questions, they become the tool-evaluation checklist. *See also* The honest scorecard.

Sound decision. A well-*made*, assured decision — complete, consistent, governed, grounded, and on the record — whatever its outcome. The discipline grades the making, not the luck. *See also* The five assurance properties; Resulting; Calibrate, never re-grade.

Super-additivity (the assembly bonus). The claim that a group, organized well, decides better than its best individual member — the wisdom-of-crowds intuition applied to a board of agents. The book pre-registered this claim for machines and watched it fail across seven regimes; the mechanism of the failure was miscalibration under aggregation. *See also* Miscalibration under aggregation; The boundary of the claim; Pre-registration.

Worse-of-two rule. The rule that a decision's level is set by the worse of its two axes — severity and reversibility — never by their average. A modest-severity decision through a one-way door is graded by the door. The single most important line in the leveling specification, because it catches the reversible-looking call that quietly closes a door it can never reopen. *See also* Severity; Reversibility; Decision Assurance Level.

REFERENCES

This section consolidates the sources named in the per-chapter notes into one reference list, organized by type so a reader — or a future contributor to the Decision Assurance Body of Knowledge — can find the spine of the argument at a glance. Foundational and scholarly works come first, then the standards and professional bodies the discipline inherits from, then the laws that are turning assurance from an option into an obligation, then the cases the book uses as evidence, and finally the author's own pre-registered research. Case and news citations are anchored to a primary source wherever one exists, because those are the claims a skeptic tests first. Page ranges, editions, and any remaining bibliographic detail are completed at copyedit; nothing here states a fact the chapters did not already establish.

A. Foundational and scholarly sources

Bainbridge, Lisanne. "Ironies of Automation." *Automatica* 19, no. 6 (1983): 775–779.

Bezos, Jeff. *Letter to Shareholders.* Amazon.com, 2015. (The one-way / two-way door, or "Type 1 / Type 2," decision framing.)

Brier, Glenn W. "Verification of Forecasts Expressed in Terms of Probability." *Monthly Weather Review* 78, no. 1 (1950): 1–3.

Cotrugli, Benedetto. *Della mercatura et del mercante perfetto.* Manuscript 1458; first printed 1573. (An early description of double-entry method, predating Pacioli's printed account.)

Deming, W. Edwards. *Out of the Crisis.* Cambridge, MA: MIT Center for Advanced Engineering Study, 1982. ("Cease dependence on inspection… build quality in.")

Duke, Annie. *Thinking in Bets: Making Smarter Decisions When You Don't Have All the Facts.* New York: Portfolio, 2018. (Resulting; decision quality versus outcome quality.)

Fama, Eugene F., and Michael C. Jensen. "Separation of Ownership and Control." *Journal of Law and Economics* 26, no. 2 (1983): 301–325. (Decision management versus decision control.)

Feynman, Richard P. "Cargo Cult Science." Caltech commencement address, 1974; reprinted in *Surely You're Joking, Mr. Feynman!* New York: W. W. Norton, 1985. ("You must not fool yourself — and you are the easiest person to fool.")

Flyvbjerg, Bent. "From Nobel Prize to Project Management: Getting Risks Right." *Project Management Journal* 37, no. 3 (2006): 5–15. (Reference-class forecasting; endorsed by the American Planning Association, 2005.)

Gabor, Dennis. *Inventing the Future.* London: Secker & Warburg, 1963. ("The future cannot be predicted, but futures can be invented.")

Garvin, David A. *Learning in Action: A Guide to Putting the Learning Organization to Work.* Boston: Harvard Business School Press, 2000. (The after-action review carried into business.)

Gawande, Atul. *The Checklist Manifesto: How to Get Things Right.* New York: Metropolitan Books, 2009. (The Boeing Model 299 / B-17 checklist origin.)

Gorry, G. Anthony, and Michael S. Scott Morton. "A Framework for Management Information Systems." *Sloan Management Review* 13, no. 1 (1971). (Decision support systems.)

Grieves, Michael. Digital-twin concept articulated in product-lifecycle-management work, circa 2002; the term "digital twin" later credited to John Vickers at NASA, circa 2010.

Herndon, Thomas, Michael Ash, and Robert Pollin. "Does High Public Debt Consistently Stifle Economic Growth? A Critique of Reinhart and Rogoff." *Cambridge Journal of Economics* 38, no. 2 (2014): 257–279. (The spreadsheet error identified, 2013.)

Kahn, Herman. Scenario work at the RAND Corporation, 1950s–1960s. (The intellectual root of scenario planning.)

Kahneman, Daniel. *Thinking, Fast and Slow.* New York: Farrar, Straus and Giroux, 2011. (The outside view, ch. 23.)

Kahneman, Daniel, and Amos Tversky. "Intuitive Prediction: Biases and Corrective Procedures." *TIMS Studies in Management Science* 12 (1979). (The outside view.)

Kelley, James E., and Morgan R. Walker. The critical path method (CPM), 1957. (With PERT, the project-scheduling ancestor of the decision dependency graph.)

Link, Edwin A. The "Blue Box" flight trainer, patented 1931; adopted by the U.S. Army Air Corps in 1934. (The first widely used flight simulator.)

Mintzberg, Henry. "Patterns in Strategy Formation." *Management Science* 24, no. 9 (1978): 934–948.

Mintzberg, Henry, and James A. Waters. "Of Strategies, Deliberate and Emergent." *Strategic Management Journal* 6, no. 3 (1985): 257–272.

Pacioli, Luca. *Summa de Arithmetica, Geometria, Proportioni et Proportionalità.* Venice, 1494. (The first printed description of double-entry bookkeeping.)

Page, Scott E. *The Difference: How the Power of Diversity Creates Better Groups, Firms, Schools, and Societies.* Princeton: Princeton University Press, 2007. ("Diversity trumps ability.")

Reinhart, Carmen M., and Kenneth S. Rogoff. "Growth in a Time of Debt." *American Economic Review* 100, no. 2 (2010): 573–578.

Ries, Eric. *The Lean Startup.* New York: Crown Business, 2011. (The minimum viable product; term coined by Frank Robinson, 2001.)

Rittel, Horst W. J., and Melvin M. Webber. "Dilemmas in a General Theory of Planning." *Policy Sciences* 4, no. 2 (1973): 155–169. (Wicked problems.)

Robinson, John B. "Energy Backcasting: A Proposed Method of Policy Analysis." *Energy Policy* 10, no. 4 (1982): 337–344.

Rockart, John F. "Chief Executives Define Their Own Data Needs." *Harvard Business Review* 57, no. 2 (1979). (And the executive information systems that followed.)

Santayana, George. *The Life of Reason,* vol. 1, *Reason in Common Sense.* New York: Charles Scribner's Sons, 1905. ("Those who cannot remember the past are condemned to repeat it.")

Surowiecki, James. *The Wisdom of Crowds.* New York: Doubleday, 2004.

Tetlock, Philip E. *Expert Political Judgment: How Good Is It? How Can We Know?* Princeton: Princeton University Press, 2005.

Tetlock, Philip E., and Dan Gardner. *Superforecasting: The Art and Science of Prediction.* New York: Crown, 2015.

Ulam, Stanislaw, John von Neumann, and Nicholas Metropolis. The Monte Carlo method, Los Alamos, 1940s. (Sampling an answer too tangled to compute directly.)

Wack, Pierre. "Scenarios: Uncharted Waters Ahead." *Harvard Business Review* 63, no. 5 (1985). (Royal Dutch/Shell scenario planning, with Ted Newland; readiness for the 1973 oil shock.)

Zheng, Lianmin, et al. "Judging LLM-as-a-Judge with MT-Bench and Chatbot Arena." 2023. (Position, verbosity, and self-preference biases in large-language-model judging; full citation to be completed at copyedit.)

B. Standards, frameworks, and professional bodies

RTCA / EUROCAE. *DO-178C: Software Considerations in Airborne Systems and Equipment Certification* (EUROCAE ED-12C). Software levels A–E. The system-level term *Design Assurance Level* is formalized in SAE ARP4754A (as Item Development Assurance Level); DO-178C itself uses "software level."

International Standard on Auditing (ISA) 200. (Defines *reasonable assurance.*)

International Standard on Assurance Engagements (ISAE) 3000. (Defines *reasonable* and *limited* assurance for non-financial subject matter.)

American Institute of Certified Public Accountants (AICPA). Trust Services Criteria; the WebTrust seal; Service Organization Control (SOC) 2 reporting.

ISACA. Certified Information Systems Auditor (CISA). (The information-systems audit profession organized in 1969.)

The Institute of Internal Auditors (IIA). International Professional Practices Framework; Certified Internal Auditor (CIA).

United States Department of Defense. The five pillars of Information Assurance: confidentiality, integrity, availability, authentication, non-repudiation. (Extending the older confidentiality-integrity-availability triad.)

C. Laws and regulations

General Data Protection Regulation (GDPR), Article 22. (Automated individual decision-making; the right to human intervention.)

European Union Artificial Intelligence Act. (High-risk obligations: human oversight, transparency, logging.)

United States Securities Act of 1933 and Securities Exchange Act of 1934. (The laws that hardened the financial-audit profession.)

United States banking model-risk guidance (Federal Reserve / OCC, Supervisory Letter SR 11-7). (Independent validation of decision models.)

D. Cases, incidents, and reporting

Google Bard / James Webb Space Telescope error. Demonstration answer factually wrong; Alphabet shares fell roughly 7.7 percent (about 100 billion dollars in market value), February 2023. Reported by Reuters and CNN. (Generative / large-language-model hallucination.)

Moffatt v. Air Canada, 2024 BCCRT 149 (British Columbia Civil Resolution Tribunal). The airline held responsible for its chatbot's invented bereavement-fare policy; the "separate legal entity" argument rejected.

Dutch childcare-benefits scandal (*toeslagenaffaire*). Algorithmic risk-profiling wrongly accused roughly 26,000 families of fraud; the government resigned in January 2021. (Algorithmic automated-decision failure, distinct from large-language-model hallucination.)

Michigan Integrated Data Automated System (MiDAS). Automated system wrongly accused roughly 40,000 people of unemployment fraud; a state review found an error rate near 93 percent; settlement of about 20 million dollars. (Algorithmic automated-decision failure.)

Knight Capital Group. Faulty trading-software deployment lost about 440 million dollars in roughly 45 minutes, August 1, 2012. U.S. Securities and Exchange Commission, *In re Knight Capital Americas LLC* (2013).

Equifax data breach. Exposure of personal data of roughly 147 million people, 2017, traced to an unapplied security patch. U.S. Government Accountability Office, GAO-18-559 (2018).

Theranos. Collapse of the blood-testing company for claims its technology could not support; U.S. Securities and Exchange Commission and U.S. Department of Justice actions (2018 onward). (Illustrates assurance without independent validation.)

Volkswagen "Dieselgate." Defeat-device software defeated emissions testing; U.S. Environmental Protection Agency Notice of Violation, September 18, 2015. (The fraud limit: no seal stops a determined cheat.)

Whisper (medical transcription). Reporting that an artificial-intelligence speech-transcription tool fabricated text in medical settings; Associated Press, 2024. (Level-A stakes met with Level-E assurance.)

Mata v. Avianca, Inc., S.D.N.Y. (2023). Attorneys sanctioned for filing a brief with fabricated case citations produced by a generative model.

Michael Cohen / fabricated citations. Legal filing containing nonexistent cases generated by an artificial-intelligence tool, U.S. District Court, Southern District of New York (2023–2024).

Hancock declaration, *Kohls v. Ellison* (2025). A sworn expert declaration on the dangers of artificial intelligence found to contain artificial-intelligence-hallucinated citations.

New York City "MyCity" chatbot. Municipal chatbot advised businesses in ways that would break the law; reporting by The Markup and the Associated Press, 2024.

JPMorgan "London Whale." Trading loss of roughly 6.2 billion dollars, 2012, involving a value-at-risk model reimplemented in a flawed spreadsheet. U.S. Senate Permanent Subcommittee on Investigations (2013); JPMorgan Chase Management Task Force Report.

Mars Climate Orbiter. Spacecraft lost on September 23, 1999, after a mismatch between imperial and metric units flowed through the navigation software. NASA Mars Climate Orbiter Mishap Investigation Board, Phase I Report (1999).

E. The author's research and related work

The pre-registered study at the center of this book. Hypothesis, method, and falsification criteria deposited on the Open Science Framework (OSF), DOI 10.17605/OSF.IO/ZWM3S, before the experiment ran. A dated Results Addendum reconciles the registered hypothesis against the disconfirming outcome; a cross-regime Capstone collects the seven regimes; a Tier-2 confirmatory pre-registration (a novel-catch metric on consented real-world cases) is drafted and, as of this writing, not yet posted. The formal record is Appendix B.

Related work by the author (lineage of the Emergent Intelligence program). "Compliments to the CHEF: A Cognitive and Heuristics-Based Emergent Financial Management Tool" (Cutter Consortium, 2017); "Taming the Wicked Problem of Portfolio Management" (Project Management Institute Global Congress, 2015). Cited as background to the parent theory, not as in-text sources.

HOW TO CITE THIS WORK

This short section does something most books never need to do, because most books are not trying to found a field. It states which parts of this work are **normative** (offered as the canonical reference for the discipline) and which are **illustrative**; it gives a way to **cite** the work and its terms; and it publishes a **versioned register** of the vocabulary, so the lexicon is governed from its first day exactly the way the discipline asks every decision to be governed. A body of knowledge that did not version its own terms would be preaching a discipline it declined to practice.

Normative versus illustrative

Two registers run through this book, and a contributor to the future body of knowledge should keep them apart.

The **normative** content is the discipline itself: the definitions in the Glossary, the five assurance properties, the Decision Assurance Levels and their specification in Appendix A, the seven jobs, the two ledgers, and the rule that outcomes calibrate but never re-grade. These are offered as the reference account — the part meant to be adopted, taught, and extended.

The **illustrative** content is everything used to make the discipline vivid: the cases, the named companies, the worked examples, and every specific number inside them. The cases are real and cited; the numbers inside the worked examples (the vendor migration, the customer-service agent in Appendix C, the ninety-day quarter) are illustrative and labeled as such. No reader should cite an illustrative number as a finding. The empirical findings of this work — and their limits — live in Appendix B and only there.

How to cite this work

For the book:

> *Guitarte, Andrew. Decision Assurance. Arclave, Inc., 2026. (Full subtitle and edition completed at publication.)*

For a definition or framework element, cite the lexicon and the section:

> *"Assurance Record," Decision Assurance lexicon v1.0, in Guitarte, Decision Assurance, Glossary.*

For the empirical result, cite the pre-registration directly, not the book's retelling of it:

> *Guitarte, Andrew. Decision Assurance: a pre-registered study of multi-agent decision quality. Open Science Framework, DOI 10.17605/OSF.IO/ZWM3S.*

A note on definitional priority

The discipline named here — **Decision Assurance** — and the apparatus built around it are introduced in this work. Where a term is original to it, the definition in the Glossary is the reference one, and this edition is its point of origin. Where a term is adapted from an established source (aviation's Design Assurance Levels, Jeff Bezos's one-way and two-way doors, the digital twin, the wisdom-of-crowds literature, Fama and Jensen's decision control), the source is credited in the References and the term is marked *adapted* in the register below. The distinction matters for a field that intends to hold itself to the grounding standard it preaches: a discipline born in citation should be scrupulous about what it coined and what it inherited.

This work, and the vocabulary it introduces, are offered to the field as the seed of the **Decision Assurance Body of Knowledge (DABOK)**, to be stewarded, versioned, and extended by the **Decision Assurance Council**. The author claims origination of the terms marked *coined*, and offers them for shared use, not private enclosure. A profession is built by giving its vocabulary away.

The terminology register

Status legend: **Coined** — original to this work; **Adapted** — built on a credited source, named in the References; **Applied** — an established term used in its established sense. Inherited terms used only in passing are credited in the References rather than listed here. Every entry is at lexicon version 0.1 and will carry a version and a change history once the Decision Assurance Council assumes stewardship.

Term	**Status**	**Built on / source**
Decision Assurance	Coined	— (the discipline named in this work)
Decision Control	Applied	Fama & Jensen, 1983
The five assurance properties (complete, consistent, governed, grounded, on the record)	Coined	—
Sound decision (made-well, not outcome-judged)	Coined	relation to "resulting," Duke 2018
Decision Assurance Level (DAL), A–E	Adapted	aviation Design Assurance Level, DO-178C
Severity × reversibility, worse-of-two rule	Coined	reversibility framing from Bezos, 2015
One-way / two-way door	Applied	Bezos, 2015

The dial; default-up rule; re-leveling	Coined	—
Assurance Record	Coined	double-entry precedent, Pacioli 1494
Outcome Ledger	Coined	double-entry precedent, Pacioli 1494
Calibrate, never re-grade	Coined	anti-"resulting," Duke 2018
Decision Dependency Graph	Coined	CPM/PERT and dependency-graph lineage
Digital twin (of a decision system)	Adapted	Grieves c.2002; NASA c.2010
Decision Flight Simulator (forward / outcome / backward)	Coined	flight-simulator analogy; Monte Carlo; backcasting
Rehearse, do not predict	Coined	Gabor 1963; Eisenhower 1957
The seven jobs	Coined	—
Decision auditor; Certified Decision Auditor (CDA)	Coined	audit-profession precedent (CPA, CIA, CISA)
Decision Audit; Seal of Assurance	Coined	assurance-engagement precedent, ISAE 3000
Reasonable versus limited assurance	Applied	ISA 200; ISAE 3000
Minimum viable practice	Adapted	minimum viable product, Robinson 2001 / Ries 2011
Governance as code	Adapted	policy as code / infrastructure as code
Append a correction (immutable ≠ uncorrectable)	Coined	erratum/record-keeping practice

Decision debt	Coined	technical-debt analogy
Decision ledger	Coined	general-ledger analogy
Decision Assurance Body of Knowledge (DABOK)	Coined	body-of-knowledge precedent (PMBOK, BABOK)
Decision Assurance Council	Coined	—
Arclave	Coined	— (the author's tool)
arclave (verb), "to lock a decision so it holds; to make it provable"	Coined	this work (verb form of the Arclave tool / the discipline)
Emergent Intelligence	Coined	the author's parent theory

The verb register

A discipline enters the language the day its name becomes a verb. This work records that step deliberately and on the record, so the coinage is dated and attributable from its first published use.

> ***arclave*** */ˈɑːrk.leɪv/ · verb to lock a decision so it holds — to make it provable. "We arclaved the cloud-migration call before sign-off."*

The verb is **coined here**, from the act the discipline performs: to *arclave* a decision is to assure it — to capture, ground, level, seal, and record it — so that it holds afterward and can be proven. By convention, the proper noun **Arclave** stays capitalized as the name of the tool; the verb **arclave** is written lowercase, the way a coined word enters ordinary use.

A note on lineage, for the reader who hears the echo. The verb sits one slant rhyme from *archive*, and the resemblance runs deeper than sound. *Archive* descends from the Greek *arkhē* — beginning, origin, and *rule* — by way of the *arkheion*, the magistrate's house where a state's records were kept and interpreted: authority and the record, bound in one word. *Arclave* carries that lineage forward and adds the one thing the archive never held — the Latin *clavis*, the key (the same key inside *conclave*, the room that locks). So the two are siblings from the two great record-keeping tongues, divided by a single task. An archive is the room where records rest; to *arclave* is to turn the key on a decision so it holds, and can be proven. One keeps the past; the other locks the present.

How the body of knowledge inherits this

The path from a book to a body of knowledge is the same path this book prescribes for any maturing practice: start small, prove it, then govern it. This edition is the minimum viable practice of the discipline's own documentation — one complete account, set down so the field can begin. The Decision Assurance Council's first task is the one the discipline assigns to every owner of a growing practice: take the rules out of a single author's head and into a governed, versioned, independently stewarded record. When that happens, this register stops being a back-matter table in one person's book and becomes the first entry in the lexicon of a profession. That is the intended trajectory, and it is offered, not asserted.

INDEX

A note on this index. Each locator is a page number; primary definitions are shown in **bold**, and consecutive pages appear as ranges. Cross-references (*See / See also*) point to the preferred entry.

A

B

C

D

P

Q

R

S

ABOUT THE AUTHOR

Andrew "Drew" Guitarte builds bridges. For twenty-five years he connected the two worlds most organizations keep apart: the technology and the business, the system and the control, the architect's blueprint and the auditor's evidence. Decision Assurance is where those bridges meet.

His career runs, fittingly, from one audit house toward another. It opened at Price Waterhouse, the global accounting and audit firm, and returned to the profession years later when he served as a Director at Protiviti, the global risk and internal-audit consulting firm. Between those two points came fifteen years at Wells Fargo, where he rose to Enterprise Business Architect and finished on the bank's Recovery and Resolution team, reporting under the Office of the Chief Financial Officer — the regulatory resolution planning, the "living wills," that the Dodd-Frank Act required of the largest banks after the 2008 crisis. He served at IBM Global Services along the way, and today leads Enduraman Corporation, his own management-consulting firm.

He works where assurance is not optional: banking and finance, Fortune 100 manufacturing, wholesale healthcare distribution, the airline industry, and government. He is a long-standing member of ISACA and a former education partner for its San Francisco chapter, a Subject Matter Expert in SAP LeanIX enterprise architecture, and the originator of Emergent Intelligence and the Pluggable Economy. He is a doctoral candidate in Business Administration and holds a Master of Science in

Computer Science, majoring in Artificial Intelligence, and teaches as an adjunct professor at Golden Gate University. Drew runs marathons with his wife.

www.ingramcontent.com/pod-product-compliance
Lightning Source LLC
LaVergne TN
LVHW102010210926
848483LV00015B/452/J